WATER MANAGEMENT IN THE ENGLISH LANDSCAPE

WATER MANAGEMENT IN THE ENGLISH LANDSCAPE
Field, Marsh and Meadow

Edited by
HADRIAN COOK AND TOM WILLIAMSON

EDINBURGH
University Press

To the drowners, drainers and marshmen of yesteryear.

© in this edition Edinburgh University Press, 1999
Copyright in the individual contributions is retained by the authors

Edinburgh University Press
22 George Square, Edinburgh

Typeset in Goudy Old Style
by Pioneer Associates, Perthshire, and
Printed and bound in Great Britain by
MPG Books, Bodmin, Cornwall

A CIP record for this book is available from the British Library

ISBN 1 85331 206 1

The right of the contributors to be identified as authors of this work has been asserted in accordance with the Copyright, Designs and Patents Act 1988.

Contents

Contributors		vii
Preface		ix
Abbreviations		xi
1.	Introduction: landscape, environment and history HADRIAN COOK AND TOM WILLIAMSON	1
2.	Soil and water management: principles and purposes HADRIAN COOK	15
3.	The drainage of arable land in medieval England DAVID HALL	28
4.	Post-medieval field drainage TOM WILLIAMSON	41
5.	Arable land drainage in the nineteenth century A. D. M. PHILLIPS	53
6.	Wetland soils DAVID DENT	73
7.	Hydrological management in reclaimed wetlands HADRIAN COOK	84
8.	Romano-British reclamation of coastal wetlands STEPHEN RIPPON	101

Contents

9. Medieval reclamation of marsh and fen — 122
   ROBERT SILVESTER

10. Post-medieval drainage of marsh and fen — 141
    CHRISTOPHER TAYLOR

11. Water meadows: their form, operation and plant ecology — 157
    ROGER CUTTING AND IAN CUMMINGS

12. The development of water meadows in the southern counties — 179
    JOSEPH BETTEY

13. Inappropriate technology? The history of 'floating' in the North and East of England — 196
    SUSANNA WADE MARTINS AND TOM WILLIAMSON

14. Historical changes in the nature conservation interest of the Fens of Cambridgeshire — 210
    CHRISTOPHER NEWBOLD

15. Water management systems: drainage and conservation — 227
    JOHN SHEAIL

References — 244

Index — 266

Contributors

JOSEPH BETTEY, formerly of the University of Bristol

HADRIAN COOK, Wye College, University of London

IAN CUMMINGS, City College, Norwich

ROGER CUTTING, City College, Norwich

DAVID DENT, School of Environmental Sciences, University of East Anglia, Norwich

DAVID HALL, Cambridge

CHRISTOPHER NEWBOLD, English Nature, Peterborough

A. D. M. PHILLIPS, Department of Geography, University of Keele

STEPHEN RIPPON, School of Geography and Archaeology, University of Exeter

JOHN SHEAIL, Institute of Terrestrial Ecology, Huntingdon

ROBERT SILVESTER, Clwyd-Powys Archaeological Trust

CHRISTOPHER TAYLOR, formerly of the Royal Commission on Historical Monuments of England

SUSANNA WADE MARTINS, Centre of East Anglian Studies, University of East Anglia, Norwich

TOM WILLIAMSON, Centre of East Anglian Studies, University of East Anglia, Norwich

Preface

This book is about the history of agricultural water management in England, and its effects upon the landscape. It is, of necessity, a multi-disciplinary work, containing contributions from leading experts in a wide range of academic fields: hydrologists, environmental scientists, historians, archaeologists and others. The subject of historic water management may at first appear an obscure or antiquarian one, but control of water has played a key, and much neglected, role in the development of agriculture in England. More importantly, the surviving remains of past management systems, and their associated habitats, form a vital element in the countryside: an understanding of their history is essential for all custodians of the natural, as well as of the archaeological, heritage. Water management systems represent one fascinating example of the complex and dynamic relationship between human societies and the natural world. A greater appreciation of the nature of that relationship is important, not only for understanding the past, but for facing the challenges of the future. This volume is one small contribution to that end.

We would like to thank Dr C. Paul Burnham, Dr Ian Cummings, Dr David Dent and Dr Jonathan Mitchley for assistance with the editorial process; Mark Penn, who assisted with field-work; Phillip Judge, who drew many of the figures; Peter Martin of Britford for his advice, co-operation and enthusiasm during field work; and Bridget Cue, for secretarial assistance. We would also like to thank our families, who have tolerated our obsession with ancient water management over many years.

Finally, we would like to thank the following institutions for permission to reproduce illustrations: Cambridge University Committee for Aerial Photography (Figures 10.1 and 11.2); English Heritage (Figure 3.3); the Soil Survey and Land Research Centre of Cranfield University (Figure 2.2); the Ministry of Defence (Figure 3.1); the Rural History Centre, Reading University (Figure 12.1); and the Spalding Gentlemen's Society (Figure 14.1).

H. C. and T. W.

Abbreviations used in references in text

DRO	Dorset Record Office
ESRO	East Suffolk Record Office
MAFF	Ministry of Agriculture, Fisheries and Food
Northants CRO	Northamptonshire County Record Office
NRA	National Rivers Authority
PD	Parliamentary Debates
PRO	Public Record Office
RSNC	Royal Society for Nature Conservation
RSPB	Royal Society for the Protection of Birds
WRO	Wiltshire Record Office
WSRO	West Suffolk Record Office
WWF	Wessex Water Authority

CHAPTER 1

Introduction: landscape, environment and history

HADRIAN COOK AND TOM WILLIAMSON

This volume explores a number of aspects of agricultural history, all concerned with the management of water, which are of key importance to people working in a wide range of fields – historians, archaeologists, ecologists and environmental managers. Although some of the matters discussed here have already been studied in depth by academics, they have never before been considered together, as a collection of inter-related practices. A full understanding of the various processes – social, economic, biological and soil-related – which underlie these practices is beyond the reach of individual scholars: hence the need for a jointly written volume. At one level, therefore, this book is an attempt to bridge an interdisciplinary gap. At another, however, it can be read as an essay in 'green history', answering the challenge set by a number of recent writers (Chase, 1992; Sheail, 1993).

With the world's population growing rapidly, the management of water resources has become an issue of increasing political importance. World-wide, it is likely that 80 per cent of water used by human societies is employed in irrigation, and although water is theoretically a renewable resource there are finite limits to the rate of abstraction at any one time. Water is essential for plant growth, yet excess water also causes problems for species ill-adapted to waterlogged conditions. In a rain-soaked island like Britain, where mean annual precipitation everywhere exceeds evapotranspiration, agriculture has always been possible without irrigation: it was the removal of excess water that has always been the most important water management practice. When irrigation did finally develop in England, probably in the early seventeenth century, it was carried out not to ameliorate a soil water deficit in arable land but – unusually, when examined in a global context – to boost the productivity of grassland. There are thus few parallels between the history of water management in England, and that in the 'hydraulic civilisations' of the ancient Near East. Nevertheless, control of water has played a key part in the development of English agriculture over the long term, and a diverse range of practices can usefully be grouped into three main categories.

Firstly, many low-lying areas of England, especially on coasts and around estuaries, are permanently waterlogged and/or subject to periodic inundation, by salt or fresh water. Much of this land is inherently productive, if excess water can be removed from it and the risk of flooding reduced. Since at least Roman-British times attempts have thus been made at reclamation, in order to extend the areas of food-producing land. Reclamation usually involved, immediately or ultimately, the raising of banks against riverine or marine flooding. Within embanked areas, the natural patterns of creek and levees, and the peat and mineral alluvium, were manipulated and ultimately (to varying degrees) transformed by various forms of agricultural use: new patterns of soil and topography were created, new landscape and habitats formed. In some places reclamation was followed by arable conversion, in others by use as pasture; in many cases prevailing land use has fluctuated over time (sometimes to a considerable extent) in response to both economic factors, especially cereal prices, and environmental factors, such as variations in climate and relative land/sea levels.

Secondly, a great deal of land in England lying above mean high tide level is occupied by soils which are, because of their texture, poorly draining – principally those formed in clays. At its worst, poor drainage can prevent the land in question from being used for arable agriculture, even though other factors – climate, markets – might encourage such use. Where it does not actually prevent arable use, waterlogging can considerably reduce the yields of the principal cereal crops. Waterlogging effects productivity in another way. Certain kinds of root crops, particularly turnips, which were crucial in the enhancement of soil fertility in the pre-industrial period, cannot successfully be grown in damp soils. For all these reasons, farmers from prehistoric times have instituted systems of field-scale drainage.

These two main forms of water management overlap to some extent, as we shall see. They are certainly very different from our third category which, as already intimated, involved not the removal of water but its deliberate application, in irrigation schemes designed to enhance the growth of grass at certain times of the year. In part, this practice was simply directed at increasing the production of wool,

Introduction: landscape, environment and history

meat or other animal products. But equally important were its effects upon arable agriculture. Increased stocking densities meant greater volumes of manure, at a time when alternative methods of enhancing soil fertility were unavailable, or in limited supply. Indeed, irrigation schemes were mainly adopted in districts in which arable farming was of primary importance, rather than in those devoted exclusively to livestock farming.

All these forms of water management have, of course, received some attention from historians and others in the past. It is, however, arguable that their overall importance has been underestimated. Agricultural historians have, for example, been much more interested in changes in cropping patterns, and in the enhancement of nutrients entering the soil which resulted from these, in explaining improvements in productivity in the medieval and post-medieval periods than in any changes effected by drainage (e.g. Campbell and Overton, 1991). It is therefore important to remember that the majority of farmland in the main arable zones of southern and eastern England benefits from improved drainage; and that some of the most productive farmland in the country occupies areas of reclaimed fen and marsh. The history of agricultural water management in England, although less dramatic perhaps than in some ancient civilisations, is thus of central importance in understanding the development of agriculture, landscape, and the wider economy, over the long term.

But this subject is also important in another, more practical way. The various practices outlined in this book constituted significant, often far-reaching, modifications of the natural environment. The conversion of the vast peat Fenland of East Anglia in the post-medieval period to productive arable land amounted to an environmental disaster on a grand scale: a complex ecosystem, already extensively modified by a sophisticated collection of management practices (grazing, reed-cutting etc.), was replaced by a landscape of arable fields and wasting peat. Yet many of the changes described in this book – like others effected to the landscape over the period since the Neolithic – have in fact served to increase habitat diversity, producing a range of new environments. Not all reclamation of primary wetlands thus produced endless arable prairies. Where coastal silt soils were reclaimed, as on Halvergate in Norfolk, grazing frequently predominated: a dense network of freshwater dykes, and extensive areas of rich pasture, sustained a varied ecology. The irrigation of water meadows likewise had profound effects upon flora and fauna, and such meadows are now, albeit mostly in abandoned relict form, of considerable conservation importance. During the twentieth century, however, agricultural change, coupled with urban and industrial development – including insensitive and expensive land drainage and flood defence schemes – have caused catastrophic damage to historic water management systems. Even where wilful damage is not manifest, neglect has often caused deterioration, and serious loss of habitat value.

An understanding of early water management systems is not only important for nature conservation. Such systems have often left physical traces that are of considerable archaeological significance. Thus the ridge and furrow used by medieval farmers to drain their fields, preserved under turf as a consequence of later land use change, still covers many square kilometres, especially in the Midlands, although with the increasing intensity of agriculture it is a vanishing resource. The complex earthworks of irrigated water meadows, still a characteristic feature of the valleys of the southern chalklands in spite of widespread destruction since the Second World War, are perhaps now under less threat, but are nevertheless vulnerable to insensitive management.

Past water management systems have thus left many traces on the landscape, and have moulded the semi-natural environment in many ways (Cook, 1994). Those responsible for managing both the natural and the cultural heritage need to be aware of the character, and significance, of these processes and techniques.

— Contexts —

The rise to prominence of the 'sustainable development' debate has provided a key forum for the discussion of the interactions between human societies and 'the environment' (e.g. Redclift 1994). Current research into 'Indigenous Technical Knowledge' (ITK) in regions as far apart as Mexico and Nepal will be familiar to some readers: the effects of the 'green revolution' are being called into question, on both economic and environmental grounds. To some extent, the history of water management in England can be read in a similar way. Until relatively recently, controlling water to enhance agricultural production does not seem to have had any widespread detrimental effects upon the environment. Marshland drainage and meadow irrigation made only a limited impact on the wider water environment and were, in this respect, in marked contrast to the present situation: the expensive flood-defences, river regulation, deep drainage and agro-chemical and sediment loading of watercourses which are the consequences of today's enhanced agricultural production. Past water management systems thus appear to have been relatively sustainable, and to have contributed in many ways to environmental diversity.

Yet the concept of 'traditional', sustainable systems coming into conflict with 'modern', unsustainable ones is nevertheless problematic. Apart from the fact that (even quite early) programmes of reclamation could have negative environmental consequences – as the example of the East Anglian Fenland demonstrates – such a model tends to suppress the extent of development and change: the idea of 'tradition' often involves a suppression of historical complexity, whereas in reality English farming systems underwent many changes over time.

These changes were motivated by a number of factors. As improved drainage, wetland reclamation, and water meadow irrigation were all aimed at raising food production, through increasing productivity per acre and/or extending the cultivated area, it might seem obvious that the innovation and diffusion of new systems and practices were principally related to population growth. This is true to some extent – and most clearly, perhaps, in the history of wetland reclamation. Yet it is important to remember that, in complex market-based economies like those of medieval and post-medieval England, farmers responded to market pressures, not directly to demography. Periods of demographic growth not only raised prices, encouraging investment in drainage, reclamation or irrigation systems: they also usually lowered wage levels. This was particularly important given that many of the practices described in this volume were cheap in terms of materials, but expensive in terms of labour. More important, however, is the fact that the relationship between demography and technical change did not always follow such a straightforward pattern. Both irrigation, and the underdrainage of arable land, spread rapidly at a time (the late seventeenth and early eighteenth centuries) of demographic stagnation, when prices were relatively low. Farmers attempted to maintain their incomes by maximising production, through adapting their local environment in ways that maximised its particular advantages *vis-à-vis* those enjoyed by other farms, in other localities.

The relationship between innovation, uptake, and demographic growth is also complicated by the regionally diverse nature of the pre-industrial, and the early industrial, economy. Thus some districts of post-medieval England – principally those with light soils, in areas of sandy heathland or chalk Wolds and Downlands – specialised in arable, sheep-corn systems of farming, in which thin easily-leached arable land was kept in heart by the systematic night-folding of flocks of sheep, grazed by day on the downs and heaths (Thirsk, 1987). In such farming systems, the introduction of irrigated water meadows made particularly good sense, for it raised the number of sheep available for the fold at critical times of the year. In other farming systems, such as those devoted to dairying, irrigation might be less appealing as an investment. In short, management systems responded in complex ways to demographic growth, and indeed to other economic stimuli, and these responses were also highly regional in character.

Variations in markets, labour supply and the economy were not the only factors determining the adoption of the various management practices described here. Institutional factors were also important. Some forms of water management involved, of necessity, the active co-operation of large numbers of individuals, at local or regional level. Others led to conflicts of interest which had to be resolved through particular forms of jurisdiction. Organisational and institutional structures are thus of key importance in the history of water management. Indeed, because of the common interest in sharing water, draining land and protecting from flood, laws concerned with water have a long history (Newson, 1992). The character of institutional arrangements – that is, patterns of legal, territorial and tenurial organisation – is, of course, intimately related to patterns of economic organisation. But in practice the two do not always develop hand in hand, the former often displaying a measure of inertia which can have a determining influence upon the development of the latter. In England the classic example of this is, perhaps, open-field agriculture. This developed in late Saxon times, doubtless for sound social, economic and political reasons. But it was incredibly long-lived, persisting in its Midland heartlands into the eighteenth and nineteenth centuries. It thus continued to have a determining influence on many farming activities (including water management practices) within a society very different from that which gave it birth.

To economic and institutional factors, we need to add the difficult issues of technology transfer and innovation. Although some of the practices outlined in this volume are relatively straightforward, others are more complex and are unlikely to have been invented repeatedly. Unfortunately, in many cases we simply do not know – and will never know – the point at which a method was invented, or the pattern of its subsequent diffusion: this is true not only of changes in the remote past, but even in some cases of post-medieval innovations. But whatever the precise character of invention and, in particular, of diffusion – whether from farm to farm, or via large landowners (private or institutional), through written texts or by oral recommendation – the actual uptake of an innovation generally depended on perceived economic benefit, mediated by the limits imposed by organisational structures; although, as we shall see in the case of meadow irrigation, adoption could also be motivated by more complex social and ideological considerations.

— THE SOCIAL AND ECONOMIC BACKGROUND —

We can now look briefly at the long-term social and economic developments that underpinned the evolution of water management practices in England. As a result of several decades of research by landscape archaeologists and others, it is now clear that pre-medieval population levels were much higher than historians once thought (Taylor, 1983). By Roman times population densities were probably comparable to those of Domesday England; this, coupled with natural falls in relative sea levels, led to the extensive reclamations of coastal wetlands described by Rippon in Chapter 8. Conversely, it was the subsequent demographic and social decline of the late Roman and early post-Roman period, coupled with a relative rise in water levels, that led to the abandonment of most of these reclaimed areas.

By late Saxon times, with population rising again and relative water levels stable or falling, a further phase of colonisation and reclamation began, principally affecting areas of estuarine clays and silts, rather than peat fens, which for the most part continued to be exploited for summer grazing, cut for marsh hay and reeds, and harvested for a range of wild resources. This time, however, as sea levels turned yet again in the twelfth and thirteenth centuries, there was no significant abandonment of reclaimed coastal lands, but rather an intensification of embanking and flood-defence activity. Embanked areas might be used for either pasture or arable – the extent of medieval ploughland in areas like Romney Marsh is surprising (Reeves, 1995).

We can only speculate about the character of the institutional frameworks that underpinned the Romano-British colonisation of wetlands. In the medieval period, however, it is clear that communities gradually combined to undertake flood-defence and drainage work, sometimes leading to the development of regional structures of organisation – like the twenty-four 'sworn men' or 'jurats' responsible for maintaining the walls and embankments on Romney Marsh against inroads from the sea or tidal rivers, and for the upkeep of drainage dykes. This group was operating from at least the thirteenth century, raising a tax to pay for necessary works and able to force landowners to repair deficiencies: 'giving judgement as to the necessary repairs in the banks and walls, and distraining offenders' (Purseglove, 1988, p. 40). By the later thirteenth century Courts and Commissions of Sewers had developed in various wetland areas (a sewer in this context is simply a drainage channel, covered or uncovered). Their responsibilities were firstly to preserve low-lying land from permanent inundation, and secondly to maintain river navigation. The first such body was established in Lincolnshire in 1258, charged with maintaining dykes, walls and bridges in Fenland (Darby, 1940, p. 155).

Other medieval institutions also had an important role in regulating wetland drainage schemes, not least the Benedictine monasteries which were the dominant landholders in many areas of marsh and fen. They instituted major reclamation schemes in the Walland Marshes, the East Anglian Fens, and the Somerset Levels (see Chapter 9). Large permanent, international institutions like these were well placed to make long-term investments, and to import techniques developed in other areas, although their role in medieval reclamation should not be overplayed. In part their activities loom larger in the historical record than those of peasant communities because they are so much better documented.

Regional systems of water management thus existed in low-lying coastal wetlands from early medieval times. More local flood alleviation measures – affecting more restricted areas of low-lying marsh and fen – were carried out by manorial courts. These bodies oversaw all basic agricultural practices in medieval England and represented, in modern terms, 'multifunctional agencies', controlling land and resource allocation and arbitrating in disputes, as well as forming the principal means by which the wealth and labour of local communities were expropriated by the feudal elite. But these institutions were not only concerned with the maintenance of drainage and flood defences in areas of low ground: they were also concerned with the maintenance of land drainage more generally, in the forms of ditches and gutters, and frequently presented individuals for causing nuisance to their neighbours by failing to maintain watercourses. Where communal organisation of agriculture was most pervasive – in the Midland areas of England, in which open-field agriculture was most developed (see Chapter 3) – the intimate involvement of manorial courts in such matters was particularly necessary, because the lands of proprietors took the form of small, intermingled strips. Outside such areas, communal systems of organisation were sometimes less well-developed, but they existed everywhere, and of necessity, in a society in which the majority of individuals held some land, and in which in most districts a farm of 30 acres (12 ha) might be considered large.

Introduction: landscape, environment and history

The demographic collapse of the fourteenth and fifteenth centuries, and the period of subsequent recovery up to c.1660, were accompanied by fundamental changes in social and economic organisation, too complex to be discussed in detail here (Overton, 1996). The average size of farms grew, and agriculture became more regionally specialised: England developed as a complex mosaic of farming regions, with many areas of heavy clay soil, even in the east of England, being put down to grass. Large capital farms steadily, if slowly, replaced small-scale producers.

In areas of reclaimed wetland, the demographic collapse in late medieval times seldom led to permanent abandonment, in spite of climatic deterioration and continuing sea-level change. Where, as in the Marshland district of Norfolk or Romney Marsh, arable land use had been extensive in the early Middle Ages there was now a marked shift from arable to pasture. But this was not so much the result of the collapse in the demand for cereals as a manifestation of the increasingly specialised nature of farming: Romney in particular continued to be predominantly a pasture district even when demographic growth took off again in the course of the sixteenth century. In most coastal wetlands, where sea defences needed to be maintained, the principal bodies responsible continued to be the various Commissions of Sewers. The Bill of Sewers of 1531 regularised the formation of Commissions, and remained the chief statute on the subject for three centuries (the number of Commissions grew steadily: by the early twentieth century there were no less than forty-nine in existence). Other Commissions – like the Sea Breach Commission, established in the Norfolk Broads following the disastrous floods of 1608/9 – were also important in maintaining low-lying land from inundation (Williamson, 1997, pp. 48–56). All such bodies, however, were principally concerned with sea defences or arterial watercourses, and paid only limited attention to local, field-scale drainage works, usually when poor maintenance threatened more general inudation. Local drainage matters continued to come under the supervision of manorial courts. Indeed, these continued to be an important factor in agrarian life generally, and in the management of water in particular, well into the post-medieval period. They played a key role in the early development of water meadow irrigation (see Chapter 12), which first appeared and spread in the early seventeenth century. The courts were a key instrument by which the principal landowners and manorial lord could override the opposition of commoners to irrigation schemes, as well as organisational structures which could be used to implement them.

Yet by the early seventeenth century the systems of communal exploitation represented by manorial courts more usually served to hinder than to encourage changes in water management systems. Thus the persistence of open-field agriculture in the Midland counties of England served, to some extent, to retard the adoption there of new forms of field drainage. From the late seventeenth century farmers in many parts of England began to install bush drains in their fields – trenches cut into the ploughsoil, filled with wood and stones. These were a considerable improvement on the various forms of surface drainage – such as ridge and furrow – which had previously been employed, but they were difficult (although not impossible) to install where land lay in intermingled strips.

In the course of the seventeenth and eighteenth centuries, however, manorial courts and communal control of land were in retreat, as the numbers of small owners, and the numbers of farmers, steadily declined, and as more and more land was amassed by large landed estates. New water management schemes were championed by large landowners and capitalist farmers, and these increasingly came into conflict with customary, 'traditional' systems of management. Legal changes – most notably, the General Drainage Act of 1600 – encouraged the involvement of outside investors in regional drainage schemes. The most dramatic example was the draining of the East Anglian Fens in the seventeenth century. Like most other areas of peat fen (and unlike areas of coastal marsh) these had largely survived through the Middle Ages in an unembanked state. Although regarded as an unproductive 'waste' by government, large landowners and outside investors, the area was intensively exploited for reeds, sedge, marsh hay, peat, fish, alder wood, and wildfowl by local commoners: primary wetlands like these are highly productive ecosystems. Unrest, especially in the 1630s, initially delayed the implementation of drainage schemes and the extension of the arable frontier, but in the later seventeenth century enclosure and reclamation destroyed much of this semi-natural habitat, and also a way of life which had, over many hundreds of years, developed with it (Hills, 1967). Ironically, the success of the 'Adventurers' was short-lived: contraction and wastage of the peat caused repeated flooding, and (coupled with a down-turn in agricultural prices in the late seventeenth century due to a period of demographic stagnation) an abandonment of arable here. Although the original environment was not re-established, by the early eighteenth century much of the reclaimed land consisted of waterlogged pastures (Darby, 1983, p. 106).

Introduction: landscape, environment and history

The Fens was not the only area where wet commons were expropriated, enclosed, and 'improved' in the course of the seventeenth and eighteenth centuries. Stokesby in the Norfolk Broads was one of many parishes where capitalism and custom collided. Here, in 1725, 'a great many poor People, both Men and Women, in a tumultuous Manner, threw down a new Mill, and diverse Gates and Fences upon the Marsh', later claiming that 'they did it for Recovery of their Right, the Marsh being common till a certain Gentleman had taken it away by fencing it in' (Malcolmson, 1981, p. 127). Enclosure and reclamation intensified in such areas in the period after 1750, when population growth resumed and agricultural prices rose, peaking in the boom years of the Napoleonic Wars (1793–1815). Field-scale drainage was steadily improved, as numerous new Drainage Commissions – able to supervise field-scale as well as 'arterial' improvements – were established by private parliamentary act, or by parliamentary enclosure awards.

This was the period of the conventional 'agricultural revolution': between 1750 and 1840 the productivity of English agriculture increased by a factor of 3.5 (Allen, 1994, p. 102), through both an improvement in yields per acre and an extension of the area under cultivation. Open fields and commons – including common fens and marshes – largely disappeared; and new crops and rotations (featuring turnips and clover) were widely adopted, in order to enhance soil fertility. Many of the new techniques were pioneered by large landowners and their tenants, although even small owner-occupiers often invested considerable sums in improvement, spurred on both by high prices and by falling labour costs. Agricultural writers and the central government's Board of Agriculture offered vocal encouragement.

The eighteenth and nineteenth centuries also saw a fundamental reorganisation of England's agricultural geography. Large areas of poorly-draining soils in the West and, in particular, in the Midlands, which had formerly been under arable cultivation, were put down to grass – often following large-scale parliamentary enclosure – in a continuation of a process which had been going on since late medieval times, thus preserving thousands of square kilometres of ridge and furrow in relict form. At the same time, arable land use expanded in the East, as vast areas of former grassland were ploughed. Indeed, in terms of water management systems, the rate of change during the 'agricultural revolution' period was startling. The late eighteenth and nineteenth centuries saw the widespread adoption of underdrainage across the claylands of eastern England, intense drainage of wetlands (especially in the East Anglian Fenlands), and the diffusion of water meadow irrigation into areas well beyond its southern and western heartlands.

In terms of conventional historiography, 'agricultural revolution' was followed by 'high farming' – the high investment, high input/output farming system of the middle decades of the nineteenth century (Wade Martins, 1995, pp. 101–12; Thompson, 1968). Artificial fertilisers came into widespread use; imported oilcake supplemented pasture and fodder crops; and thus the fertility of the farm was no longer primarily dependent upon what could be produced within its own boundaries, but rested instead upon a range of commodities imported from the manufacturing sector – or from other parts of the globe. Such changes were truly revolutionary, but in terms of water management systems the high farming period was characterised as much by refinements to existing practices as by the adoption of entirely novel ones. Encouraged in part by a gradual rise in labour costs, landowners invested in permanent, and more technologically sophisticated, versions of existing improvements. Thus bush drains – cheap in materials but of short duration – were steadily replaced by earthenware tile pipes, which spread rapidly in the period after 1840 and brought underdrainage into many areas previously unaffected by the improvement (see Chapter 5). In the wetlands, steady improvements in windmill technology were accompanied by the development of steam drainage, through which the arabilisation of the East Anglian fens was finally achieved. But this period also saw the beginning of the end of some long-established forms of management. In particular, by the middle of the nineteenth century the widespread adoption of new root crops, and the growing availability of artificial fertilisers, were beginning to threaten the viability of water meadows.

The period of high farming was brought to an abrupt end in the late 1870s by the onset of a severe agricultural depression, caused by large-scale imports of grain and, later, refrigerated meat, from the New World. Depression continued, with only short periods of upturn, until the start of the Second World War in 1939 (Mingay, 1994, pp. 194–244). Throughout the late nineteenth and early twentieth centuries, therefore, investment in agriculture was at a very low level, and large areas of arable land reverted to pasture. Some of the more marginal areas of reclaimed peat wetland were abandoned – especially in the Norfolk Broads; the installation of underdrains virtually came to an end and existing systems deteriorated; and, encouraged in part by the widespread availability of artificial fertilisers, the management of water meadows steadily declined. What was bad news for farmers and landowners, however, was probably on the whole good news for nature conservation: wildlife flourished in a landscape less intensively farmed. All this, however, was to change rapidly from the 1940s.

— POST-WAR LANDSCAPE AND HABITAT CHANGE —

Wartime blockade saw a rapid extension in the area under arable cultivation which, supported first by national government and subsequently by EEC policy, continued into the 1990s (see Chapter 15). In terms of nature conservation, this recovery in agricultural fortunes had almost entirely negative effects. Old grassland usually exhibits a considerable degree of biodiversity, especially in marshes and water-meadows, where dykes and channels support a wide variety of flora and fauna. When arable replaces grassland, species diversity is rapidly reduced: arable systems are managed to achieve a monoculture within a small area; 'weeds' are rigorously controlled by the application of herbicides. Coastal marshes have suffered badly in these respects, especially in the east of England. Large areas of Romney Marsh and Halvergate were ploughed in the 1960s and 1970s. Since the 1930s, almost 60 per cent of marshes have been lost in Greater London, the Thames Estuary and adjacent coastal areas of Kent and Essex (RSPB, 1992). Arable intensification in such environments usually involves the installation of underdrains, the deepening of drainage dykes, improved vehicular access and enhanced flood-protection measures – all of which have serious effects upon habitat value and landscape.

The remaining areas of 'primary' wetland in England – principally found on peat rather than silt or clay soils – also fared badly in the period after the Second World War. 'Primary' is, of course, something of a misnomer: these environments had long been intensively exploited by local communities, and their vegetation and hydrology extensively modified. Neglect of traditional management can thus be as damaging as intentional destruction through drainage: the most extensive areas of remaining wetlands, in the Norfolk Broads, have thus suffered considerable damage in the course of the twentieth century as regular mowing for reed, saw sedge and marsh hay have declined, leading to the encroachment of open fen by scrub and carr. Eutrophication arising from effluents and agro-chemicals, coupled with physical disturbance through drainage schemes and visitor pressure, have also served to diminish this rich mosaic of habitats and landscapes.

The loss of wetlands has had other important negative effects. Because anaerobic conditions favour the preservation of organic materials, waterlogged environments are an archaeological resource of immense importance. Organic materials such as wood and textiles are likely to be preserved here, as well as pollen and other natural materials that are able to provide information about the wider environment and its exploitation. Many of the most important excavations of the last two or three decades have been in such locations: the prehistoric trackways in the Somerset Levels, the Bronze Age settlement at Flag Fen. Deep drainage and consequent de-watering of wetlands, together with large-scale commercial peat extraction, thus have important archaeological implications.

In many, perhaps most contexts, the preservation of the natural and the cultural heritage go hand in hand: the partial ploughing of the Halvergate marshes in Broadland in the 1960s and 1970s, for example, not only damaged a nationally important dyke flora but also an ancient landscape, as a pattern of dykes established in early medieval times was filled and levelled. Ploughing of old grassland in the Midlands not only destroys an often herb-rich pasture but also eradicates remaining areas of ridge and furrow. Archaeologists and ecologists increasingly need to make common cause, and we hope that this volume may, in some small way, reinforce the growing links between them.

In many respects the tide of agricultural destruction has turned. Government grant aid to underdrain potential arable land is a fraction of the support offered twenty years ago. The recent rise of water protection zones and schemes for the protection, restoration and enhancement of wetland and riverine habitats (such as Countryside Stewardship) has made water protection and aquatic habitat conservation measures major factors in countryside management (Cook, 1993). The greatest threat to some important areas of the countryside can now come, not from wilful destruction, but from poor management. But best practice depends ultimately on an understanding of the character of landscapes and habitats, of how they came into existence, and of how they developed over time. We hope that this volume will provide a better understanding of one important aspect of the history of the countryside.

Chapter 2

Soil and water management: principles and purposes

Hadrian Cook

— Why drain? —

In contrast to ancient 'hydraulic' civilisations of the Middle East and elsewhere, unirrigated agriculture in Britain has always been possible. It is usually seasonal excess, rather than shortage, which has created problems; and where soils are prone to waterlogging, improved drainage brings a number of benefits. Firstly, it enables the soil to warm up more rapidly in the spring: it takes about three to four times the heat energy to increase a unit volume of saturated soil through one degree Celsius compared with one of dry soil. Furthermore, in poorly-drained soils, more water has to be removed by evaporation, and this serves to cool the surface, as heat energy is lost through the latent heat of evaporation. Reducing this 'thermal inertia' is critical for germination and early crop development; denser, water-retentive clay soils are frequently described as 'cold'.

Secondly, better drainage improves aeration, not only increasing the diffusion of atmospheric oxygen into the soil, but also encouraging the diffusion out of carbon dioxide arising from soil biomass respiration. Diffusion rates for oxygen are a little higher than those for carbon dioxide in both air and water; however, the rates of diffusion for both is about four orders of magnitude less in water than in air. Increasing the water content of the soil thus increases the amount of carbon dioxide present, and decreases the amount of oxygen.

Thirdly, drainage improves soil aggregation by causing shrinkage and cracking, and – by encouraging the activity of the soil biota through de-watering – provides channels for root development and fissures for drainage water. It also improves a soil's 'trafficability'; that is, its ability to permit the movement of a vehicle over the land surface without experiencing compaction, and hence damage to the soil structure, which can further impede drainage (Reeve and Fausey, 1974).

There are a range of additional benefits. Drainage deepens the soil available to the rooting system by increasing the depth of 'freeboard' drainage above the watertable; that is, the distance between the soil surface and the watertable – effectively defining the 'unsaturated' or 'vadose' zone, in which soil aeration is essential for crop development. It also prevents the loss of the soil resource itself through wind and water erosion following the disaggregation of topsoil structure; and reduces the 'heaving' caused by freeze and thaw action, which can be a serious problem in saturated soil, and particularly disruptive to root systems.

The significance of soil profile waterlogging for the diffusion in of atmospheric oxygen and diffusion out of carbon dioxide is especially critical. When a soil contains free oxygen, whether gaseous in the soil air or dissolved in soil water, aerobic respiration (essential to the root system), mineralisation, the conversion of ammonium to nitrate and the oxidation of sulphide to sulphate, can all occur. On the other hand, waterlogging for more than 48 hours is likely to cause all the soil oxygen to be consumed by aerobic respiration because oxygen diffusion rates in wet soils are some 10,000 times slower than those for drained soils and so, without replenishment, the soil biomass soon becomes oxygen deficient (Correl and Weller, 1989). The consequences for most plants (rice and reeds excepted) are serious. The roots will cease to respire, and the microbial population switches from aerobic to anaerobic respiration pathways. This has a number of serious implications, including the evolution of organic and inorganic phytotoxic substances (such as ethylene) and the loss of dissolved nitrate in the soil to the gaseous phase through microbial denitrification. Denitrification is the reduction of nitrate (NO_3) to nitrous oxide gas (N_2O) and ultimately dinotrogen (N_2). A saturated soil will experience a series of reduction reactions as anaerobic processes 'switch in' where there is decomposable organic matter. The reduction-oxidation, or 'redox', potential of water is an important variable in controlling chemical reactions in aquatic systems. The redox potential is a measure of electron availability in a solution and hence quantifies the degree of electrochemical reduction in soils.

Reduction is when:

1. Oxygen is given up.
2. Hydrogen is gained (hydrogenation).
3. Electron gain occurs.

Reduction-oxidation potential is conveniently measured in millivolts (mV) on the Eh scale, using an inert platinum electrode. Where free, dissolved oxygen is present ('aerobic conditions'), values are generally in the range +400 to +700mv; but when reducing (e.g. 'anaerobic') conditions pertain the range tends to be between +400 and −400mv. In a waterlogged soil, where organic substrates are oxidised, and the redox potential falls (i.e., moves towards or into the negative range), electron gain occurs. Below-ground reduction processes are complex, but may be summarised as follows:

Reduction processes (in the presence of organic matter represented by 'CH_2O') (after Correl and Weller, 1989).

E_{h7} approx +400mV: Manganese (iv) reduction
$2MnO_2 + \text{'}CH_2O\text{'} + 4H_3O + \rightarrow CO_2 + 2Mn^{2+} + 7H_2O$

E_{h7} approx +300mV: Denitrification
$NO_3^- + \text{'}CH_2O\text{'} + H_3O^+ \rightarrow CO_2 + \frac{1}{2}(N_2O) + 2(\frac{1}{2}H_2O)$

E_{h7} approx −200mV: Iron reduction
$4Fe(OH)_3 + \text{'}CH_2O\text{'} \rightarrow 4Fe(OH)_2 + CO_2 + 3H_2O$

E_{h7} approx −220mV: Sulphate reduction
$SO_4^{2-} + 2\text{'}CH_2O\text{'} + 2H_3O^+ \rightarrow H_2S + 2CO_2 + 4H_2O$

$E_{h7} \leq -260$mv: Methanogenesis
$2\text{'}CH_2O\text{'} \rightarrow CO_2 + CH_4$

---

note E_{h7} is redox potential at pH7.

Eh is both temperature- and pH-dependent; the latter being the strongest influence. For example, the reaction which reduces nitrate will fall in Eh potential by 59mV per unit increase of pH.

The development of reduced species of both ferrous iron (Fe^{2+}) and Manganese (Mn^{2+}) can adversely affect plant growth (although the latter is only toxic to some species in soils of low pH). It is the presence of reduced iron that produces the characteristic 'gleying' of certain subsoils which are subject to prolonged waterlogging. Where the effects are seasonal, and generally due to the vertical undulations of a watertable, a characteristic mottling results. This displays green, grey and bluish pockets of reduced iron species (Fe^{2+}) interspersed with orange, yellow and brownish soil containing ferric iron (Fe^{3+}). The latter is a common feature of former root channels and worm burrows which favour the transport of oxygen from the atmosphere.

Organic compounds may also result from anaerobic conditions through fermentation. One group of substances, termed volatile fatty acids, may be produced by the fermentation of carbohydrates arising from crop residues (White, 1997). These are toxic to plants, especially young seedlings; but due to their low molecular weights, are lost to the atmosphere as CO_2, CH_4, and H_2 by the action of anaerobes. Nitrogen is released through protein decomposition as NH_4^+, which can be the dominant form of nitrogen in grassland soils.

Low molecular weights are produced during the early stages of waterlogging, and their concentrations in heavier-textured soils tends to be higher; methane is a particular example. Most such hydrocarbons are not harmful to crops, but propylene and butene have small inhibitory effects upon root development, whilst ethylene (C_2H_4) can cause serious damage. This is evolved in anaerobic soils through fermentation of sugar fungi *Mucor* sp. and in concentrations above 1mg/1 it can seriously inhibit root development (Russell, 1973, Ch. 25).

The above discussion may cause difficulties for those readers without some grounding in natural science. However, the overall effects of soil waterlogging may be summarised as follows. Reduction rids the soil of free oxygen and causes the evolution of anaerobic conditions. This leads to a loss of the root system's ability to respire after about 48 hours. In addition, the soil experiences a loss of soil nitrate (essential for crop growth) due to microbial breakdown to gaseous forms of nitrogen, and also a significant reduction in its ability to warm up, enhance growth and trigger germination. Waterlogging leads to the evolution of a range of organic and inorganic substances which are toxic to plant growth. Apart from such essentially chemical effects, poor drainage makes a soil prone to structural damage by reducing its trafficability; and once it is damaged, a soil is prone to erosion, through the destruction of its natural aggregation.

— Climatic factors —

Not all soils, of course, are in equal need of drainage, and to assess the situations in which drainage is most required we must first turn to climatic considerations. Table 2.1 shows the rainfall, mean water balances, excess winter rainfall and period of 'field capacity' for six regions representative of case studies in this book. The regions are arranged in order of increasing rainfall, and the figures refer to the period 1941–70.

Mean rainfall figures stand only as a guide; rainfall is very variable in both space and time, and water management systems have to cope with extremes of rainfall, and to include strategies for dealing with flooding. In all the regions selected, there is an average excess of rainfall throughout the year; but during the summer, in all but one region, there is an excess of potential transpiration over rainfall for the months April to September. The balances shown are both for the entire year (January to December) and for the agronomically important period April to September; the former increases ten-fold moving from the top to the bottom of the table, while the summer deficit decreases, becoming positive in the case of Exmoor.

TABLE 2.1 Climate and field capacity conditions for selected sites in southern England

Area	Annual average rainfall and range (mm)	Annual balance (mm)	Balance April to Sept. (mm)	Median excess winter rainfall (mm)	Median field capacity Start	Median field capacity End
Cambridge	574, 510–730	+51	−159	130	10 Dec.	27 Mar.
Norwich	623, 520–710	+93	−152	160	11 Dec.	29 Mar.
Romney Marsh	683, 540–850	+120	−158	200	20 Nov.	8 Apr.
Salisbury	799, 620–1,000	+288	−81	315	26 Oct.	21 Apr.
Taunton	865, 650–1220	+342	−66	360	22 Oct.	20 Apr.
Exmoor	1265, 880–2000	+792	+103	725	25 Aug.	20 May.

Source: Smith and Trafford, 1976

Excess winter rainfall is one approximation of the severity of a drainage problem (Castle et al., 1984), and represents the period during which the drains are likely to be working, excepting periods of very heavy rainfall during the remainder of the year. It is calculated on the basis of the time the soil is at 'field capacity' (see below) or wetter; the median figure derived from the rainfall distribution is quoted for each example. Essentially, this figure becomes greater with increasing rainfall, because the rate of evapotranspiration from west to east across Britain is relatively conservative.

— Soil profile considerations —

Soil and water management depends not only upon climatic considerations, but also on fundamental properties of the soil. Physical characteristics are immediately apparent, and it is descriptions such as 'sand', 'silt', 'clay', 'loam' or 'peat' which provide the basis for soil classification and management. When considering soil profiles, a 'zero point' is required in calculating soil water balances, and this is taken to be 'field capacity'. The concept of 'field capacity' is problematic to many hydrologists, because soil water and water vapour are perpetually on the move over different scales. Practical agronomic considerations, however, make it essential to establish a benchmark for calculating drainage and soil water deficits for purposes of irrigation scheduling. Field capacity may be defined as the water content of the profile following heavy rainfall (i.e. saturation) and drainage for 48 hours, in the absence of both evapotranspiration and shallow watertable influences (Ward and Robinson, 1990, Ch. 5); it is supposed that after this period water movement will have virtually ceased. In reality, soil texture and relatively impermeable horizons will complicate the issues, with the more permeable soils draining more rapidly and some, typically silts, so slowly that it becomes difficult to identify when drainage has 'virtually ceased'.

During the autumn, as evapotranspiration falls, the soil profile will re-wet, eventually reaching the 'field capacity' condition. In practice, the date of this occurrence will vary from year to year so the median (i.e. mid-point of the range of dates experienced) is selected in order to characterise the expected times of arrival of field capacity for a particular region. The period between this time and the median end of field capacity conditions (Table 2.1) defines the period of winter drainage. Excluding the western hills (represented by Exmoor), this ranges from three and a half months in eastern England to six months in the Somerset levels.

Figure 2.1 shows a schematic diagram of water content profiles in a soil. It represents a heavier textured clayey soil. There are three conditions: (1) saturation: (2) 'field capacity' (when drainage has effectively ceased and (3) the point at which all the water held under physical conditions accessible to the crop is exhausted. This defines the so-called 'permanent wilt point', when wilting commences and the plants will not recover without irrigation or rainfall. Water held between conditions (1) and (2), in excess of field capacity, is liable to drain; that held between (2) and (3) is termed the 'profile available water capacity', because it was once assumed that this water was all equally available for crop uptake. While still useful, this concept (like that of field capacity) has been attacked for being too simplistic, and specifically for not considering the progressive onset of crop water stress (e.g. Hillel, 1982) which is capable of influencing the amount of water which can be taken up.

Most agricultural land requiring drainage consists of 'intermediate clay' (30–50 per cent clay) or 'heavy clay' (50–80 per cent) soils, especially in southern and eastern Britain: examples include the Beccles series soils, and others developed on the chalky boulder clays of East Anglia and adjacent regions; the Denchworth series, formed over Jurassic and Cretaceous clays in the Midlands and South; and (to a lesser extent) the Batcombe soils, formed in clay-with-flints in areas of

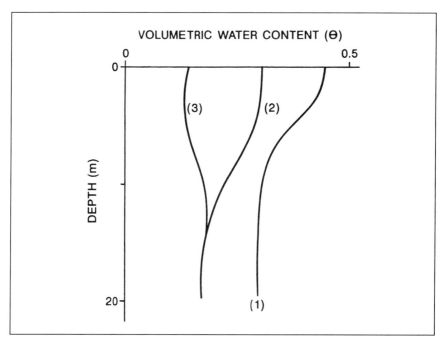

Figure 2.1 Soil water content profiles at (1) saturation, (2) field capacity and (3) permanent wilt point.

chalk Downland, especially the North Downs and the Chiltern Hills. Various low-lying fen and marsh soils also require drainage, including the Adventurer's series – deep, amorphous and semi-fibrous eutrophic fen peats – and the Newchurch and related soils, characteristic of areas of drained alluvial marsh. Many upland areas, such as Exmoor (Curtis et al., 1976) display a range of soils which require drainage, including podzols, gleys and organic 'bog' soils, as well as brown earths and complexes comprising several types in relatively small areas.

The distribution of soils requiring drainage in central and eastern Britain is shown in Figure 2.2, expressed in terms of their soil water regimes (Mackney, 1975). Soils affected by shallow groundwater (2a, 2b) include river valley alluvium, and much of the coastal alluvial and reclaimed peatlands of the Somerset Levels, the East Anglian Fens, and the east and south coasts. Boulder clay and clay vale soils largely fall into category 3a.

The importance in historical studies of soil survey mapping is that it gives a common framework in which to compare soils and their management through time. However, it should never be forgotten that economic imperatives (especially markets for produce), local tenurial customs (notably regarding land parcelling) and other social factors, in the past as today, may have a more important effect on land use patterns than the character of soils *per se*. Indeed, over the long term, areas of soils requiring drainage have been exploited in a diversity of ways, with drainage technology developing accordingly. Thus the claylands of East Anglia, now highly productive arable land, were largely under grass in the seventeenth century; areas of ridge-and furrow in grass fields in the Midlands attest the former use of these heavy soils as arable; while Romney Marsh, used as pasture throughout post-medieval times until extensive arable conversion in and after the Second World War, had large areas in tilth in the middle ages.

– Methods of soil drainage –

As the following chapters demonstrate, a wide range of methods has been employed over the centuries to drain land. The earliest and simplest method was to surround a field with a ditch, a practice known from prehistoric times: notable examples include those created as part of a pastoral landscape during the second millennium BC at Fengate near Peterborough (Fowler 1983, p. 195). Field ditches are standard features of relict landscapes of prehistoric and Roman British

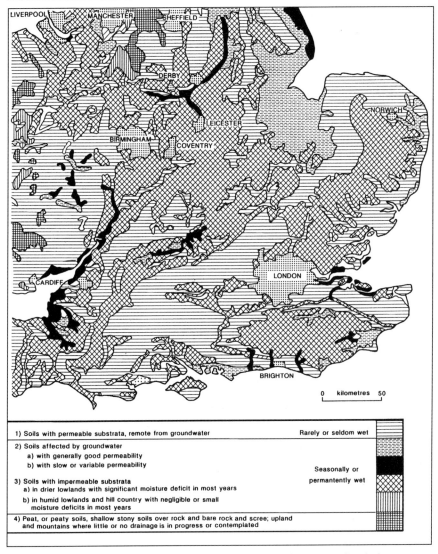

Figure 2.2 Soil water regimes in central and south-east England. (Source: after the Soil Survey of England and Wales).

date revealed by aerial photography. Today, ditches comprise integral parts of many lowland landscapes, notably 'secondary' grazing marsh and drained peatlands (see Chapter 7) but also, less noticeably, in areas of clay, where they are usually accompanied by hedges (and only intermittently fill with water). Where fields were under arable cultivation drainage was often assisted by some form of ridging or corrugation, but it was in unenclosed contexts – in the extensive unhedged arable fields of the medieval Midlands – that such methods were most developed, in the form of the broad, prominent ridges known as 'ridge and furrow'.

In post-medieval times these methods were supplemented, or replaced, by various forms of underdrainage. The earliest form of this practice involved cutting trenches through the surface of fields at intervals of 12–15m, which were filled with faggots or stones and then backfilled with soil. Later, in the nineteenth century, tiles or pipes came into use. Underdraining activity reached a peak in the middle decades of the nineteenth century, but was largely abandoned as a result of the agricultural depression of the late nineteenth century; serious interest only revived after 1940. Cylindrical ceramic tiles continued to be used, and were commonplace during land drainage schemes during the 1960s. Since then, perforated plastic pipes have become the most common method. The estimated area underdrained in 1954 was 32,000ha; the rate had risen to 100,000ha per annum by 1978, although it was falling again by the early 1980s (Trafford, 1970, 1985). The economics are complicated, but levels of grant aid have played an important part in determining the variations in rates of progress.

However accomplished, underdrainage achieves an increase in freeboard drainage by lowering the watertable while at the same time not increasing the density of open surface drains. The latter remain, however, forming the link between the field-scale and regional drainage arrangements. Underdrains are generally laid in such a way

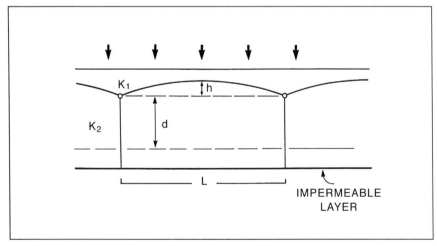

Figure 2.3 The influence of underdrainage on a shallow watertable

as to intercept natural groundwater flows, and with a gradient sufficient to convey disaggregated soil particles, which would otherwise clog them up. Originally, underdrains were installed by eye, the drainer relying on knowledge and experience. Agricultural writers in the late eighteenth and nineteenth centuries provided more detailed advice (e.g. Mitchell, 1894) and in the twentieth century, formulae inevitably emerged, most notably the 'elliptical formula' published in 1940 by S. B. Hooghoudt (Raadsma, 1974):

$L^2 = (8 K_2 d h q^{-1}) + (4 K_1 h^2 q^{-1})$
L = drain spacing in metres
K_1 = horizontal saturated hydraulic conductivity above the level of the drains (metres day^{-1})
K_2 = horizontal saturated hydraulic conductivity below the level of the drains (metres day^{-1})
d = thickness of the 'equivalent layer' in metres (i.e. applies to K_2)
h = the height of the groundwater table above the plane midway between two drains in metres
q = drainage discharge or drainage coefficient (metres day^{-1})

This steady-state formula tells us that reduction of saturated hydraulic conductivity (i.e. the ability of the soil to transmit water horizontally in the saturated state), stated height of the groundwater mid-way between drains and thickness of the 'equivalent layer' would all increase the required spacing of drains; conversely, increasing the discharge capability of the system would decrease the required drain spacing. Experience has shown that a watertable of minimum height 0.5m should be maintained. Discharge criteria used to calculate drain spacing (Castle et al., 1984) are typically 7mm for grass, 8mm for arable land, and 11mm for land in horticultural use, to accommodate heavy rainfall events.

Mole drainage, achieved simply by dragging a bullet-shaped mole plough through the soil, was developed in the eighteenth century and is today a widespread practice in clay soils, with drains spaced at 5m or less and with a depth of between 0.5 and 0.6m (Parkinson, 1995). In contrast to pipe underdrainage, the life of a mole drain is usually only a few years, but the method works well enough on 'stiff' soils in which the proportion of clay exceeds 30 per cent.

Drainage affects the use and productivity of individual arable fields in a wide variety of ways, as will become apparent in the course of this volume. But it also has a wider impact on the environment, in the way that it affects the volume of water flowing through streams and rivers. These wider catchment effects may include reduction of flash flooding, and an increase in the volume of arterial drainage discharges. There is also an increase in the loading of nitrates (Green, 1979b) and other agro-chemicals. Detailed catchment responses depend upon the extent and location of underdrainage, local cropping patterns and soil composition.

— THE BENEFITS OF DRAINAGE —

Field drainage is clearly beneficial: it unquestionably leads to an improvement in yields, although the extent varies from crop to crop (Trafford, 1970). Marshall (1959) reports research that demonstrates the advantage of drainage on yields of potatoes, rye, sugar beet, kale, celery and ryegrass on fen peat soils, such as the Adventurer's Series. In general, yields increased towards the optimum as watertables fell towards 0.8m, although celery and potatoes produced maximum yields at 0.6m; presumably these water-stress sensitive crops benefited from shallower watertables. To assess the yield benefits of underdrainage, MAFF studies (Rose and Armstrong, 1992) have utilised a 'Yield Benefit Index (BI)':

$$BI (\%) = \frac{\text{yield drained} - \text{yield undrained} \times 100\%}{\text{yield drained}}$$

This is typically between 10 and 15 per cent for arable crops. Increases in grass yield are also achieved by drainage: the BI values reported for drained grass, as well as gains through increases in trafficability and reduction of poaching, are considerable and the overall levels of animal production are often increased by the same order as those for cereals (Rose and Armstrong, 1992). Castle et al. (1984) report a study in Wiltshire where grassland yields increased from an average of 13.3 to 15.3 tonnes ha^{-1} yr^{-1}. Draining produced an increase of 148 per cent in spring yields, and nitrogen uptake exceeded 100 per cent for the undrained plots.

Grass growth, however, is not only limited by an excess but also by a shortage of water. Whereas, until very recently, irrigation was not used on arable crops, it was widely practised on certain kinds of grassland in the post-medieval period. Regrowth is inhibited by a soil water deficit of 60mm or above (MAFF, 1966). Dry matter yields have been found to increase by 27 per cent with (modern) irrigation, rising to 83 per cent when nitrogen fertiliser was also applied (Moore, 1966). One estimate for early 'bedwork' watermeadow grass production put the yield at 50 percent above that for unirrigated land (Moon and Green, 1940).

It was only in the period after the Second World War that detailed research was undertaken in order to quantify fully the costs and benefits of drainage improvements. In practice, these are difficult to quantify, for a wide range of factors needs to be taken into account, including a comparison of the costs of the capital investment with the deterioration of the system over time. Following the immediate post-war enthusiasm for virtually unlimited land drainage, such analyses – considering all the inputs and 'externalities' (including grant-aid input, amenity and conservation losses, actual yield benefits and the actual rate of land conversion) – cast doubt on the maintenance of the practice on purely economic grounds (e.g. Bowers, 1983; Morris, 1989). It is certainly clear that the increasing scale of land drainage in the second half of the twentieth century has had a major hydrological and ecological impact: as early as 1979 Green (1979a) was able to state:

> Field underdrainage is intended to lower the watertable, for agricultural purposes, at times where it would otherwise be too high. But it has other effects, ecological and hydrological as well as agricultural, mainly because it alters the response to rainfall of the outflow from the subsurface drains into the watercourses, as well as lowering the mean level of the soil watertable.

These themes are developed by John Sheail in the final chapter of this volume: the next three chapters explore the development of land drainage methods from medieval times until the nineteenth century.

CHAPTER 3

The drainage of arable land in medieval England

DAVID HALL

— RIDGE AND FURROW —

Old pasture fields in many parts of the country have their ground surface lying in regular patterns of parallel low ridges, commonly called 'ridge and furrow' (Figure 3.1). Pasture is now relatively rare in eastern England, but some survives in the Midlands, and a particularly large expanse of ridge and furrow can be seen at Lilbourne, Northamptonshire, where the A14 joins the M1 and M6.

Ridge and furrow displays a variety of forms, some confined to specific regions of the country, others reflecting changing agricultural practice. It is essential to be clear about the dating and origin of these differing forms, and in particular to distinguish between ridges produced after enclosure, mainly in the nineteenth century, from those that can be related to earlier open-field cultivation. Post-enclosure ridges occur in two forms, one wide and the other narrow. Examples of the former are up to 15m in width and have a different general appearance to pre-enclosure ridges. Their chief characteristics are their straightness, and the way in which they lie parallel to at least one modern field-hedge. Furrows are picked up by a 'headland furrow' going around the edge of the modern field. Most examples survive in the central and western Midlands, e.g. north of Banbury in the Sheninton area (Oxon). Several farmers from Warwickshire have reported that their grandfathers still ploughed in ridges like these until about 1920. Dated examples occur in Northamptonshire at Naseby (Grid reference SP 688 787), in an area enclosed in 1820 (illustrated in Hall, 1982, p. 11); at King's Cliffe (TL 014 993), where a mid nineteenth-century date is proved by the fact that the area in question was a medieval wooded coppice, part of Rockingham Forest, until about 1860 (marked on Bryant's 1825 county map but removed

The drainage of arable land in medieval England

Figure 3.1 Aerial photograph of typical high-backed ridge and furrow, Padbury, Buckinghamshire. Note the well-developed 'heads' at the ends of the lands in the centre of the picture. The parish was enclosed in 1795 (© Crown Copyright/MOD.)

before the Ordnance Survey First Edition 1:10,560 in 1885: nearby woods were stubbed up in 1860). Mead (1954) refers to Rowley Regis in Staffordshire, where an early nineteenth-century lease agreement states that every field was to be gathered into ridges 5 yards (4.5m) wide when fallow. Nearly all the ridge and furrow surviving on clay soils in Kent in the 1940s was of this straight and late type, confined to present-day field boundaries (Kain and Mead, 1976).

Another type of nineteenth-century ploughing has left very narrow ridges, only a metre or so wide. These are rare in the Midlands; examples lie over part of the earthworks of the deserted villages of Onley and Sulby in Northamptonshire (SP 520 715 and SP 653 815: Beresford and St Joseph, 1979, p. 39). A more extensive area survives at Wappenham (SP 636 443) in fields that were still part of a medieval woodland coppice in 1761 but which were converted to agriculture by 1822 (Baker, 1822, p. 725). They are confined to the present field boundaries and lie parallel to straight hedges. Such narrow ridges were more extensively used in the north of England, for instance in the Manchester region. Examples occur in south Lancashire around Denton (SJ 903 951 and SJ 887 941: Reddish). At nearby Alkrington ridges lie on the peat surface of White Moss, an area not drained and brought into agriculture until c.1840 (Hall et al., 1995, pp. 112-13). In the undulating terrains of the north, blocks of narrow ridges were fitted into the landscape wherever the soil was sufficiently planar to plough. A modern field may contain several blocks with different alignments which have no relation to existing field boundaries, other than being confined by them. Several authors refer to such 'narrow rig' ploughing techniques. Murray (1813), writing about Warwickshire, noted that 'old ridges' vary in width but that new small ridges were 2 yards (1.8m) wide. No more will be said here about these relatively late forms of ridging, the present concern being with medieval methods of land drainage.

In the past, historians took it for granted that ridge and furrow was related to medieval cultivation. Sir John Lubbock in 1892 equated ridge and furrow with former open-field husbandry, and Maitland, writing in 1897, noted 'the practice of ploughing the land into... ridges... anyone who has walked through English grass fields will know what they look like...'. Beresford studied ridge and furrow in 1948, relating ground observations to maps and written records, and demonstrated the pre-enclosure origins of much ridge and furrow. This was disputed (Kerridge, 1951; Mead, 1954; Kerridge, 1955), but only because of confusion with post-enclosure ridging of the type described above.

In many parts of the country, methods of open-field agriculture survived into the eighteenth and nineteenth centuries and are sometimes very fully recorded on maps, often accompanied by detailed written surveys. Court rolls, listing agricultural orders and infringements of those orders, give some indication of how such fields were worked. Many full accounts of them are available (Seebohm, 1883; Roberts, 1973; Hall 1995, pp. 1-8). In the Midland region their layout was as follows. Here a 'farm' was called 'yardland' and consisted of about 25 acres (10ha) of land (the amount varied within each township), lying not in a block, but scattered in long, narrow strips (called 'lands') throughout the township, no two strips lying together. The strips originally probably covered an area of a quarter or half an acre, but later often had average dimensions of about 8 yards (7m) by 200 yards (190m), thus being nearer to a third of an acre. The ends of most lands are curved, so that the whole took the shape of a very elongated, mirror-image of an 'S'. This shape seems to have developed over the years, resulting from the ploughman's tendency first to draw out to the left when performing a turning circle to the right (Maitland, 1897, p. 440; Eyre, 1955). Groups of lands lying parallel to each other were called 'furlongs', and were identified by names. Early in the Middle Ages most of the available ground in a Midland township was ploughed in furlongs; only river meadows liable to flooding, wet areas along the side of brooks where springs emerged, and areas with steep and rough ground, were left unploughed. As much as 90 per cent of a township area was used for arable.

For the purposes of cropping, furlongs were grouped together in large open areas called 'fields'. Typically there would be two or three fields in a township, and each year one would be left fallow. It is for this reason that a single holding was scattered uniformly throughout the arable lands; if it were not, then all or most of a farmer's land might lie in a fallow field in one year and there would be no produce to live on. The fallow field, although producing no crop, was valuable as rough grazing for the village animals. It was essential to have a large, continuous block of land so that the common herd could be kept together and animals prevented from wandering on to the cultivated area.

As well ridging a land, the action of the plough moved small quantities in the direction of motion, towards the ends. Over the years small heaps formed at each end. They were called 'heads' and are first recorded in the thirteenth century (Figure 3.2). The soil piled up at the end of the lands does not become much flattened by modern ploughing, so that it forms a long smooth bank along the edge of the former furlong. Wherever a furlong met another, either orientated in the same direction, or at right angles – that is, at the headlands or joints – a bank of soil will survive. Such soilbanks are visible over most of lowland England. They are not, however, proof that such areas were ridged, only that furlongs once existed. Soilbanks would form whether lands were ridged or not, the key mechanism being that soil accumulates where ploughs turned, slightly more soil being deposited each time than was taken away on the return.

— THE FUNCTION OF RIDGE AND FURROW —

Intermixed, dispersed holdings of small strips in the open field were primarily a tenurial arrangement, but the ground needed to be cultivated by a method suited to times when underdraining was not employed. Ploughs made a single furrow and turned soil towards the right as the ploughteam moved forward. Each land was ploughed clockwise, beginning in the middle and going round and round until the outside was reached. The centre would have two cuts piled against each other, and the two final outer cuts left furrows, so the action of the plough moved soil towards the centre (Hall, 1972, p. 55) (Figure 3.2). Repetition of this process formed permanent ridges at the centre, and furrows at the edge, of each strip.

Sixteenth- and seventeenth-century agricultural authors provide much information about the details of farming in open fields. Lands were not ploughed in the same direction every year, or they would

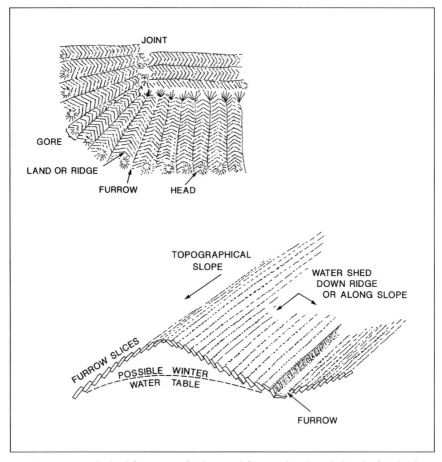

Figure 3.2 Principal features of ridge and forrow (top) and detail of individual strip (bottom), showing how ridging serves to remove water from the land.

have become very ridged, with furrows cutting deeply into the infertile subsoil. An anticlockwise motion was adopted in the fallow season. Plot (1705, p. 244), writing about Oxfordshire, calls fallowing 'a casting tilth, i.e. beginning at the out-sides of the lands, and laying the earths from the ridge at the top'. The purpose of ridging was to obtain a well-drained seed bed; Plot continues, 'ridging it up twice or thrice for every casting tilth ... beginning at the top of the land and laying the earth still upwards to the ridges, by which means both land and corn lie dryer, warmer and healthier, and the succeeding crop becomes more free from weeds'.

The furrow lying between ridges acted as a drain, as well as providing a clear boundary between the ground of each tenant. Furrows were aligned downhill, running across contours, and the local topography and natural drainage therefore determined the furlong pattern. Across the country variations are apparent: on gentle slopes furlongs can be over 1.5km long, made up of hundreds of lands lying side by side. In contrast, high undulating ground can have very complex patterns of small furlongs, with lands lying in many directions.

The steepness of ridges varies according to two main factors: soil type, and date of enclosure. Study of dated examples of ridge and furrow, preserved within pasture fields of known enclosure date, shows that there was a general trend for ridging to become steeper as time passed. Ridge and furrow found in early enclosures has a rather lower profile than examples known to have been under the plough in the eighteenth and nineteenth centuries. At Wollaston in Northamptonshire the demesne land was enclosed in two stages, in 1231 and 1583, and this contains low-profile ridges which contrast markedly with the more pronounced ridge of adjacent land ploughed until 1788 (Hall, 1982, Figs. 8 and 4). The high ridging of the later period is probably a reflection of the fashion for agricultural improvement, and represents an attempt to keep open-field tilth in good condition in the absence of underdraining.

Soil type also has some influence on ridging: it is usually higher on clay ground. Regional preference and fashion also come into play, however: towards East Anglia, where ridging was not favoured in the eighteenth century, profiles are low, even on quite heavy marly soils such as occur near Cambridge.

Although it is clear that ridging was carried out primarily to drain individual lands, it did not in itself deal with the problem of overall water management in the open fields. The alignment of ridges across the contours ensured that water was moved to the lower ends of furrows. If their ends opened on to a meadow or a stream, water would drain away naturally; but where they lay against a joint or the headland of another furlong, where there was (as already noted) a bank of soil, there could be problems, especially in nearly flat terrain. Recurrent difficulties are attested by the frequent occurrence of such names as 'water furrows furlong'. In the sixteenth and seventeenth centuries the problem was dealt with by making outlets using a trenching spade or trenching plough. Water from one furlong would pass to the next and so to a meadow or 'gutter', the usual Midland term for a natural watercourse running in the bottom of a minor valley. Maintenance of water courses is recorded in open-field orders dating from the sixteenth century, and it is likely that the same techniques were used in the Middle Ages. At Hemington, Northamptonshire, manorial courts of 1523 and 1554 thus demanded that tenants having a plough were to drain furlongs after Michaelmas so that water would run out of the lands and meadows (Northants CRO, Buccleuch 25/74 in box X889). All the householders of Lamport, in the same county, were to ensure that Peas Slade was 'scoured out from the furlong' before 11 November 1567. The implements used for such purposes were illustrated by Blith in 1652 (McDonald, 1908, p. 101). Walter of Henley, in the thirteenth century, mentions making water furrows to deliver water from wet ground (Oschinsky, 1971, p. 323).

— The distribution of ridge and furrow —

Even in the 1990s ridge and furrow is widely distributed and remains fairly common in many districts. However, it is difficult to ascertain its overall national distribution in earlier centuries, before enclosure; current survival is a complex function of what was there originally, the ploughing techniques used immediately before enclosure, and what has happened since enclosure.

The Midlands are best known for ridge and furrow and considerable quantities still survive along the border of Warwickshire and Northamptonshire, in Leicestershire and in central Buckinghamshire. A recent study of vertical aerial photographs taken in c.1990 has mapped surviving ridge and furrow in these counties, and also in Bedfordshire, Oxfordshire, south-west Lincolnshire and western Cambridgeshire (Foard and Hall, 1997, pp. 69–71) (Figure 3.3). The current distribution coincides with predominately clay soils and, except for central Buckinghamshire, with undulating scarp topography. Much the same area was mapped from 1940s RAF vertical photographs by Mead and

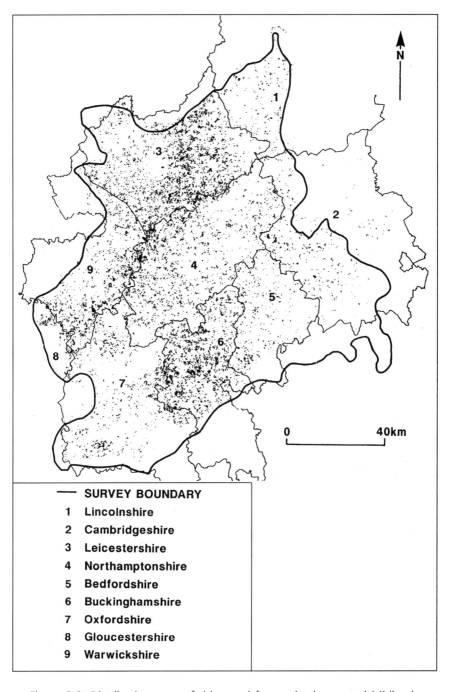

Figure 3.3 Distribution map of ridge and furrow in the central Midlands, 1990. Earthworks of this type now survive only in very limited pockets, mainly on the heavier soils: three decades ago large areas of the Midlands were carpeted with fossilised ridges (Courtesy English Heritage.)

his co-workers (Mead, 1954; Harrison et al., 1965; Kain and Mead, 1977), and it is alarming to see the amount of destruction that has occurred in less than fifty years.

Ridge and furrow coincides with the area of the 'Midland System', as defined in 1915 by Gray: that is, with so-called 'regular' open-fields of the kind described above. It is associated with a region of nucleated villages that extends in a belt far beyond the Midlands, through Yorkshire to Durham and eastern Northumberland, and south-west to Somerset. The zone has recently been mapped in detail by Roberts as part of a study of settlement types, and designated the Central Province (Roberts and Wrathmell, 1995). Ridge and furrow occurs, even today, throughout this region, there being good examples outside the central Midlands in Derbyshire (Brassington: Beresford and St Joseph, 1979, p. 133), the East Riding of Yorkshire, Durham, Gloucestershire, Wiltshire and Somerset (Aston, 1988).

The Central Province is essentially part of the lowland zone of England. In upland regions there is much rough ground and moorland that was never ploughed in the Middle Ages. Open-field systems did exist in these regions but they were, for the most part, both smaller and more complex than those of the Midlands (Baker and Butlin, 1973). The regional terminology is correspondingly different and in many places infield and outfield cultivation was practised, the outfield being left to revert to pasture for several years at intervals, in order to recover fertility. There is some evidence for ridging, and for the creation of furrows for drainage between ridges, in Lancashire and elsewhere in the North-West (Elliot, 1973, p. 47). Northumberland and Durham were similar; the term 'rig' occurs in documents, suggesting that ridging was practised and confirming the physical evidence (Butlin, 1973, p. 101). There is a reference to ridging at Guysance in Northumberland in 1567 (Butlin, 1973, p. 177). Cheshire has examples of ridge and furrow (e.g. at Bradley, SJ 506 485, (Williams, 1978), and Castleton in Caldecott (A. Mayer, personal communication 1997)). South-east of the Central Province, the Chilterns had ridge and furrow (Roden, 1973, p. 328) but the fields here were complex and the settlements dispersed, like those of the North-West. In East Anglia some ridge and furrow is known, mainly on claylands in the west (Silvester, 1989), and may once have been common (Liddiard, 1997). The details of field systems, their terminology, and the disposition of holdings within the fields is very different from the Midlands (Postgate, 1973, pp. 290–3). It is therefore apparent that ridge and furrow was once widely distributed except for South-Eastern England, the far South-West, and on high unploughable ground, mainly in the North.

In the Midlands, and elsewhere, the dimensions of ridge and furrow are as given above. In other regions there are variations in the sizes of lands. The gravel terraces of the Welland Valley north of Peterborough had lands wider than those described, being up to 22 yards (20m) in width and making, therefore, one-acre strips. Wide lands are common, too, in the Cambridge region, being frequently 11 yards (10m) across, giving lands of half an acre.

Away from the Midlands and the Central Province, the pattern of strip-holding can be appreciably different, with lands of great length and few furlongs. On the Yorkshire Wolds, at Wharram Percy, Burdale and Wharram le Street, there are many examples of lands between 500 yards and a mile (450m-1.6km) in length (Sheppard, 1973, p. 145). The whole area was divided, with lands running across the high plateaux from dale to dale (Hall, 1982, pp. 51-2). Most of the Wolds are now arable but ridge and furrow survives in a few fields and in spinneys. Lands can still sometimes be seen in arable fields in the form of soilmarks, which stretch away in gentle curves over great distances. Similar long lands occurred in the Vale of Pickering and in Holderness, Yorkshire, where they ran uninterrupted from the village to the township boundary – a distance of as much as a mile (1.6km) (Harvey, 1981).

The overall distribution of ridge and furrow is similar to the map showing the extent of open field systems, of various kinds, published by the Orwins in 1938 (p. 65). Open fields did not necessarily always have ridge and furrow, as explained, but the two were very closely related.

— THE ANTIQUITY OF RIDGE AND FURROW —

Comparisons between open-field maps, and ridge and furrow appearing on aerial photographs, were made by Beresford and St Joseph (1979, pp. 26-36) and reveal an exact correspondence. Similar correlation of ground surveys and aerial photographs was made for seventeen parishes with open-field maps (ranging in date from 1583 to 1846) in Bedfordshire, Cambridgeshire and Northamptonshire in 1981. In all cases there was exact agreement of furlong patterns (Hall, 1981, p. 25).

The earlier existence of ridge and furrow can be demonstrated in cases where earthworks can be related to particular events recorded in documents. Thus at Titchmarsh in Northamptonshire a manorial site was enlarged to create a small park by including part of an adjacent furlong in 1304: the rest of the furlong remained in cultivation as part of the parish's open fields until 1779. On the inner side of the park boundary the ridges have a low profile, but on the outside they are steeply ridged, with a new series of heads abutting the line of the park pale. These different ridges align on each side of the park boundary and clearly pass under it, neatly demonstrating that the original single furlong was ploughed in ridges before the year 1304 (Hall, 1995, p. 39). The demesne of Hall manor at Wollaston has already been referred to. Here, the earliest intake of arable occurred before 1231, when the manor-house boundary extended as far as a 'ditch dug anciently through old ploughlands of the time' (Leicester County Record Office, ID50/xii/28, p. 20). The area remains under grass and the earthworks of the ditch can be seen curving through low-profile ridges which are presumably of twelfth-century, or earlier, date (Hall, 1982, Fig. 8).

Agricultural texts and documents confirm that lands were ridged from early times. Walter of Henley, writing in the late thirteenth century, notes that the land or ridge should lie in a 'crest' (Oschinsky, 1971, p. 321). At Moreton Pinkney (Northamptonshire) in c.1250 a terrier described the parcels in several furlongs as 'ruggis' (British Library, Harl. Ch. 84 I 25). Lands at Prescott were ridged by 1433, to judge from a reference to 'iiii butts . . . called fowr rygges' (Northants CRO Th 183 m.7). Beresford (1954, pp. 89–90) quotes historical sources to show that early post-medieval observers equated ridges with land that had previously been under cultivation, like the surveyor of Whatborough, Leicestershire, in 1583, who observed ridges in pastures and noted 'theis groundes have . . . bene arrable'.

— Regions without ridge and furrow —

Outside the extensive area of England in which ridging is well attested, evidence for ridge and furrow is meagre. There are a number of reasons for this. One is that not all areas of England were characterised by open-field agriculture. In some districts, such as the Weald of Kent or central Essex, the landscape was dominated by enclosed fields, and open arable was of limited extent or absent altogether. Here drainage depended on deep field ditches, associated with perimeter hedges. These generally formed a complex network, linking together features of the human landscape, such as moats and common edges, and natural watercourses. The latter might be extensively modified in their upper reaches, in order to improve the movement of water, and as a result it is often difficult to say whether a particular watercourse represents a stream or a drainage ditch.

In many other areas, however, open fields of a kind were widespread, and the character of field draining, and the reasons for the absence of surviving ridge and furrow, are unclear. In East Anglia, as already mentioned, it has traditionally been thought that ridging was never practised. Silvester (1989) noted some examples in the west of Norfolk, and a scatter in the south and east of the county. He assumed that fields were seldom ridged in the region, the practice spreading from the Midlands only as far as west Norfolk, via the Fen islands. More recently Liddiard (1997) has identified further Norfolk examples, not all in the west, and has suggested that probably much of the county once had ridge and furrow but that it has been ploughed flat since the eighteenth century, when underdrainage was introduced. Most of Norfolk was enclosed at an early date, and this encouraged the removal of ridges when mixed, rather than pastoral, farming was practised.

Eastern Cambridgeshire, like Norfolk, has no surviving ridge and furrow, even though early maps show open-field systems with strips organised into furlongs in the normal Midland manner. There is written evidence that strips were ploughed flat in the district: Kain and Mead (1977, p. 133) quote the 1794 Board of Agriculture report stating that in the south-eastern parish of Study Camps open fields at the time of enclosure lay flat and were under drained, with no high-backed lands. In 1598 it was also stated that lands here lay flat, without ridge and furrow, as in most parts of Cambridgeshire (Kain and Mead, 1977, p. 135). It is unlikely that hollow drains were in used at that date, but most townships in south-eastern Cambridgeshire lie on gentle slopes of a sand-capped chalk ridge and have very light soils. They would probably not have needed much draining. Maps of places like Chippenham and Burwell (Cambridge CRO R/58/16/1 and 152/P5) show that strips were aligned across the contours and down the gradient, and this was probably enough to achieve the necessary drainage.

Fitzherbert described the types of plough used in different parts of the country in 1523 (reproduced by McDonald, 1908, p. 14). In Kent a two-way or turnwrest plough was used so that sods all lay in the same direction. This kind of Kentish plough was also referred to by Blith in 1652 (McDonald, 1908, p. 97). Ridge and furrow would not be formed by the use of such techniques, which explains the absence of early ridge and furrow in that county. Drainage was presumably by alignment of 'flat furrows' across gradients, as on the Cambridgeshire chalklands.

In the silt Fenlands of East Anglia, a different type of subdivided field could be found. The whole area of medieval dry land was divided into strips, not ridged up but bounded and drained by dikes; the widths commonly varied from 12 to 20m, although examples up to 50m wide also occur. They were called 'darlands', 'darlings' or 'dielings'. A block of strips was called a 'field'. In fields near to the villages strip lengths were not great, but in areas reclaimed during the twelfth and thirteenth century massive fields were created, with strips up to one mile (1.6km) in length. In most cases a 'field' consisted of a single group of strips with the same orientation, that is, corresponding to a 'furlong' in the Midland region, except that the Fenland fields covered a much greater area (Hall, 1996, Plate 13, Fig. 99). Water from ditches flanking the flat strips was picked up in drains surrounding the fields; these emptied at low tide by means of sluices running through sea defences.

From the above discussion it is evident that ridge and furrow was widespread throughout England, although little now survives except in regions with clay soils. Ridging is unquestionably a relic of open-field agriculture, dating from the Middle Ages, and was carried out primarily for seed-bed drainage, although it also acted as a tenurial demarcation between adjacent holdings. Drainage from the furrows of each furlong was arranged as far as possible so that it discharged directly into small natural watercourses. Where this was not possible, temporary channels were cut through headlands from one furlong to the next, so linking with a natural outlet. The process was managed, like other aspects of open-field agriculture, by manorial courts.

CHAPTER 4

Post-medieval field drainage

TOM WILLIAMSON

— SURFACE DRAINAGE —

For much of the post-medieval period, farmers continued to rely on various forms of surface drainage to remove water from their arable fields. In enclosed land, water draining down furrows was conducted to field ditches, usually associated with hedges: leases on heavier land often paid particular attention to the tenant's obligation to maintain these in good condition. They were usually dug out every five or ten years, during the winter, when the accompanying hedge was coppiced or laid. In unenclosed land, the flow of surface water was less tightly controlled and pools of standing water were a frequent sight in open fields.

Surface drainage was often assisted by some form of ridging. As David Hall has already emphasised, the practice of ploughing arable fields in broad ridge and furrow did not die out at the end of the medieval period. It continued in all areas of heavy soils wherever open fields were cultivated, and was even used in some enclosed contexts. By the end of the eighteenth century, however, agricultural writers were expressing disapproval of the method. Nathaniel Kent in 1793 thus described how, in the Vale of Evesham:

> The land is thrown into ridges, from ten to thirty yards wide, and raised in the middle, to an elevation of, at least, a yard above the level, which is attended with great loss, and inconvenience. The furrows very often contain water three yards wide. The headlands are thrown up in the same manner, which dams up the water in the furrows, so that it cannot get off, but rots the seed, and destroys the crop. (Kent, 1793, pp. 21–2.)

Other kinds of ridging were also employed, although a proliferation of local terms, and some interchangability of terms, sometimes makes it difficult to distinguish between them. William Folkingham, in his *Feudographica* of 1610, differentiated between two different ploughing techniques suitable for clay soils. While 'fat, strong and fertile grounds that be tough, stiffe, binding cold and wet' should be ploughed in broad ridges, 'cold and stiffe ground inclining to barrennesse' should be ploughed in 'stitches' (sometimes described as 'stetches'): a practice which, he noted, was common in East Anglia and Hertfordshire (Folkingham, 1610, p. 48). The main difference between ridge and furrow, and stitches, was that the latter were much narrower – no more than would be produced by two or three passes of the plough – and were not usually permanent. The ridges could be cut the following year, so that new ridges would be made in the old furrows or the ground ploughed flat; alternatively, the direction of ploughing could be changed in alternate years (or at intervals of several years) and the field cross-ploughed. 'Stitching' seems to have been practised principally in enclosed fields, rather than in open-field contexts, where ridge and furrow of normal form was more likely to be employed, partly because of the size and shape of the individual strips.

Land was also ploughed in permanent narrow ridges, producing the earthworks known as 'narrow rig' – that is, diminutive ridge and furrow, each ridge generally around 1–1.5m wide and less than 0.3m in height. Some land seems to have been ploughed like this in the Middle Ages – Drury (1981) reports an example from Essex – but the technique seems essentially to be a post-medieval one. Many examples of narrow rig appear in contexts which suggest short-term ploughing of relatively marginal land during the Napoleonic War period: in intakes from waste in Shropshire and Staffordshire; in areas of late eighteenth-century reclamation in the New Forest; and in various areas of chalk Downland in southern England (Taylor, 1975, pp. 143–4, 148). As Hall describes in the previous chapter, remains of narrow rig are particularly widespread in the north of England. Kent appears to describe this method in 1793:

> Another mode of draining ploughed land is, by throwing it into very small ridges of two, sometimes four, sometimes six furrows only; and provided the ground be ploughed in such a manner, as to give the furrows a free discharge, this is by no means a bad practice; because it takes off all surface water, and the land is not more difficult to occupy, and may be thrown into any other form at pleasure. (Kent, 1793, p.23.)

Narrow ridges are referred to by a number of mid nineteenth century writers as a normal aspect of husbandry (Stephens, 1851; Andrews, 1853).

Ridging in various forms thus continued to be employed by farmers in England throughout the post-medieval period. Nevertheless, it gradually lost favour in the course of the eighteenth and nineteenth centuries, and was replaced by various forms of underdrainage.

— UNDERDRAINAGE —

Underdrainage is achieved by cutting drains beneath the surface which remove water first downwards, beyond the root zone, and then laterally, away from the field. Although described by Kent in 1793 as 'the most effectual way of draining ploughed ground' (Kent, 1793, p. 23), the importance of underdrainage before the middle decades of the nineteenth century has often been underestimated by historians. Typical is the assertion of Chambers and Mingay that 'the problem of heavy land drainage remained unsolved until the introduction of cheap tile drainage in the 1840s, leaving the high-cost and inefficient farming of heavy claylands as the most obvious weakness in the progress of eighteenth century farming' (Chambers and Mingay, 1966, p. 65). In fact, although the technique was made more efficient, and was more widely adopted, with the development of earthenware pipes in the course of the nineteenth century, underdrainage was widely implemented in some areas before *c*.1840.

Before the adoption of earthenware tiles and pipes in the nineteenth century, underdrainage was usually carried out using 'bush' drains. As their name implies, these were trenches cut across fields, which were filled with brushwood and/or various other materials and then backfilled with soil. The drains either emptied directly into the ditches surrounding the field, or into a larger underground drain which did so. Field drains of this kind would commonly last between ten and fifteen years, although they could survive for longer: Arthur Young in 1804 reported some at Redenhall in Norfolk which were still serviceable after 27 years (Young, 1804a, p. 392). Caird reported that in Suffolk in the 1850s the benefits from such drains were normally 'expected to last for a 14 years' lease', and went on:

> At the beginning of a new lease the land is gone over again, the direction of the drains being now made to cross the old drains obliquely, and thus to *bleed* such as still remain open. As a long fallow is regarded

as a routine operation twice or thrice in the course of a short lease, so is draining looked upon as a matter of regular occurrence once every 14 or 16 years. (Caird, 1852, pp. 152-3.)

Bush drainage is often described as a primitive practice compared to that which was effected by tiles or pipes (Harvey, 1980, p. 72). But this is only true up to a point. Although earthenware drains lasted longer but there is no real evidence that they were much more efficient, and they were more expensive to install. It is noteworthy that in the middle of the nineteenth century prominent agriculturalists could still recommend the old method in preference to the new. Some argued that tile drains would become clogged with soil, others that that the process of renewing bush drains every fifteen years or so, and the breaking up of the soil which this entailed, helped the water to descend more rapidly after rain (Raynbird, 1849, p. 116). The agent for the Ashburnham estate in Suffolk described in 1830 how he encouraged the tenants to follow the old method of making drains, which lasted 'many years' although created at a fraction of the cost of those made with tiles. 'I have urged all the tenants to underdrain all the land requiring it, assuring them that they will themselves partake of the benefit and be reimbursed by the first crop' (ESRO HB 4/2).

In part, modern historians' neglect of bush drainage is a direct reflection of the importance they have ascribed to the later ceramic pipes, and this in turn probably results from the greater visibility of the latter in documentary sources. The various forms of tile drainage were discussed at length in the pages of the *Journal of the Royal Agricultural Society of England* in the 1830s and 1840s, and the government loans scheme instituted in the 1850s generated a substantial amount of documentary evidence that has been used by historians (see Chapter 5). Bush drains, in contrast, have left little documentary evidence. Because they were an improvement unlikely to last much beyond the term of a lease, they were seldom mentioned in tenancy agreements, being considered the concern only of the tenant, not the landlord. What we know about the practice thus comes from post-medieval agricultural texts, letters, and in particular from farming diaries, like that kept by Randall Burroughes of Wymondham in Norfolk between 1794 and 1799 (Wade Martins and Williamson, 1995).

Analysis of this latter source suggests that Burroughes was draining more than 25 acres of his farm, mainly situated on heavy boulder clay soils of the Beccles series, each year. His drains were generally spaced

Figure 4.1 Labourers installing bush drains on a Norfolk farm in the late nineteenth century. (Source: Rider Haggard, R., (1899). A Farmers Year.*)*

at intervals of 12 yards (11m) and dug to a depth of 24 or 26in (60 or 65cm). They fed into main drains dug to a depth of 28in (70cm) (Wade Martins and Williamson, 1995, p. 27). These dimensions were typical – similar, for example, to those quoted by Arthur Young in his *General Views* of Norfolk and Suffolk (Young, 1804a, pp. 389–93; Young, 1813a, pp. 172–3). Burroughes sometimes used stones to fill his drains but normally employed faggots, cut from coppices or pollards, and sometimes hedge cuttings; this was the usual practice in the old-enclosed, well-hedged counties of Essex, Hertfordshire, Norfolk and Suffolk, where underdrainage first became established. In some areas of the Midlands, in contrast, newly-enclosed from open fields in the course of the eighteenth century, faggots and hedge trimmings were often in short supply and drains were more likely to be filled largely, or entirely, with stones. Thus Marshall described the use of 'pebbles, provincially "bowlders", picked off the arable lands' to fill drains in parts of this region (Marshall, 1796, p. 141).

Whatever the mix of materials used as fill, this was usually sealed with a layer of straw before backfilling, although other materials (such as furze) were occasionally employed. Such was the importance of bush drainage in the eastern counties that the need for capping materials could have important knock-on effects upon the way in which crops were harvested. Thus the agent for the Ashburnham estate in Suffolk described in his survey of 1830 how the wheat stubble was 'left standing in the fields of considerable length having been reaped with a sickle in that manner' so that it could be used for drains (interestingly, the tithe file for Earsham in south Norfolk noted with approval that the farmers there used furze and ling for their drains, 'knowing too well the use of straw for manure to bury it in this shape') (ESRO, HB 4/2; PRO, IR 18 6336 5896).

Whatever the precise method employed, bush drainage was not cheap. Burroughes himself described it as 'an expensive improvement'. In 1795 he informs us that the 514 rods of drains put into a field called Block Close cost £9 9s 4d (at the standard rate of a halfpenny a yard), but added that this was for labour only, 'exclusive of materials and carriage amounting to more than equal that sum', suggesting a total cost of £20. Careful analysis of the figures given in this and other years suggests that underdrainage cost Burroughes around £2 per acre, more or less the figure suggested by Philip Pusey in his remarks following Henry Evans' article 'Norfolk Draining' in the *Journal of the Royal Agricultural Society* for 1845 (Wade Martins and Williamson, 1995, pp. 27-8; Evans, 1845). Kent in 1793, however, estimated a cost of £4 18s 6d per acre, the difference perhaps accounted for by the fact that he was writing with less wooded countryside in mind, so that the costs of wood (including carriage) were much higher, amounting to £2 13s 4d per acre (Kent, 1793, p. 26). Where stones instead of faggots were used as filling, the costs of might be higher still, although the drains would last longer. Kent concluded that, on cost grounds, he would chose 'either of these methods according to the ease with which the materials are obtained' (Kent, 1793, p. 29).

In the eighteenth century bush drainage was largely restricted to arable land. Meadows were sporadically improved by the installation of a few drains, to relieve waterlogging in springy corners, but 'thorough' drainage of such land appears to have been rare. On some pasture land another form of drainage was sporadically employed, especially where the soil was particularly stiff and tenacious. Turfs were removed and shallow slots cut into the ground beneath: the turf was then replaced. Such simple drains might function effectively for several years, thought Kent, but every effort had to be made to keep cattle off 'till the drains have time to settle' (Kent, 1793, p. 30).

—THE CHRONOLOGY OF BUSH DRAINAGE—

It is not entirely clear when bush drainage first became widely adopted in eastern England, in part, once again, because of ambiguities of terminology: the phrase 'bush draining' or 'underdraining' could embrace both the installation of a single drain, to relieve waterlogging in a springy field corner; and 'thorough' drainage, that is, the installation of a complete network of underdrains within a field. The practice was certainly known in the seventeenth century – Walter Blith refers to it in 1649 (Wade Martins, 1995, p. 102) – but, so far as the evidence goes, it did not become widespread until the eighteenth. In 1845 Henry Evans was informed that the practice had 'prevailed for a century and a half', in East Anglia, but Pusey in the same year suggested a shorter chronology, stating that '*for a century* it has been used generally in the large and well-farmed counties of Essex, Suffolk, and Norfolk, as well as in Hertfordshire' (my italics) (Evans, 1845; Pusey, 1845).

It is probable that thorough drainage was first widely used in Essex, in the first decades of the eighteenth century: contemporaries often referred to it as the 'Essex method' of draining. Bradley in 1727 described the practice as 'but a late invention' on the claylands of north Essex, while in the following year Salmon described how the 'cold and wet lands' of Hertfordshire had been 'greatly improved' by underdraining, an innovation introduced 'within twenty years' from Essex (Bradley, 1727, p. 23; Salmon, 1728, p. 1).

The spread of underdrainage northwards, into Norfolk and Suffolk, seems to have been gradual. There are scattered references to the practice in East Anglian farming accounts in the first half of the eighteenth century, but when a survey was made of the Blickling estate farms in Wymondham on the south Norfolk claylands in 1750 it was noted that 'both the tenants like all others in the woodland or dairy part of Norfolk are slovenly and don't endeavour to drain their land in a husbandlike manner – heavy strong land, but could be improved' (NRO, MC 3/59/252 468 X 4).

In 1786 Young described how at Crowfield in Suffolk hollow draining was 'done by the farmers, but not the twentieth part' that should have been carried out (Young, 1786, p. 196); and one of the contributors to the Rainbirds' *Suffolk Agriculture* of 1849 reported 'the statement of old farmers, who allege that sixty or seventy years ago the practice was just being introduced into the parish in which they had been brought up' (Raynbirds, 1849, p. 112).

On the other hand, Young described in the *General View* for

Norfolk in 1804 how 'draining is well established and much done' in the area between Attleborough and Hingham; while in the *General View* for Suffolk in 1813 he simply remarked that 'this most excellent practice is general on all the wet lands of the county; it is too well known to need a particular description' (Young, 1804a, p. 392; Young, 1813a, p. 172-3). Randall Burroughes of Wymondham was an ardent improver, and it is possible that he drained his land more assiduously than his neighbours in the 1790s, but a number of references in the journal indicate that he was not a trail-blazer. In 1795 he agreed to supply the drainers with their tools although 'according to custom' they should supply their own: a phrase suggesting that underdraining was a common and well-established practice in the area (Wade Martins and Williamson, 1995, p. 28).

By the early nineteenth century, certainly, underdrainage was widely practised in Norfolk, Suffolk, Hertfordshire and Essex, and the tithe files from the late 1830s, for the two East Anglian counties in particular, frequently comment on the importance of the improvement and the widespread nature of its adoption. In the north Suffolk parishes of South Elmham All Saints and St Nicholas, for example, it was said that 'there is a great spirit of improvement pervading this part of the county and that by means of underdraining which is now in very general practice the produce of these heavy lands will be very much increased' (PRO, IR 18 9736).

Negative comments on the subject also occur, but their phrasing implies that the practice was becoming normal. Those made about North Lopham in Norfolk are typical: 'good and effectual underdraining is constantly necessary and this in the case of the small owners and occupiers is sometimes neglected on account of the heavy expense' (PRO, IR 18 6069).

The available evidence thus suggests the following chronology for the development of underdraining in the prime arable areas of eastern England. The principle of making bush drains was understood from at least the seventeenth century, but the practice was first adopted in a systematic way, and on a large scale, on the boulder clays of north Essex and east Hertfordshire in the early eighteenth century. From here it spread at first gradually and then – after c.1760 – more rapidly; and by the 1790s was widely established in many parts of the eastern clays. By the 1850s, when bush drains began to be supplemented by tile pipes, underdrainage was a standard part of clayland husbandry in eastern England, and one of the contributors to the Raynbirds' survey of Suffolk agriculture in 1849 went so far as to assert that 'at the present time nearly every piece of land in the heavy land district of this county [Suffolk] has been drained, and many pieces several times' (Raynbirds, 1849, p. 115).

At first sight it might appear surprising that a practice bringing such obvious and immediate benefits should spread so slowly through these arable counties. The high costs of the improvement were probably the main factor here, coupled with gradual changes in the economics of farming. As already noted, at least half the cost of installing bush drains was in labour – for in most cases, the materials used to construct the drains were freely available on the farm. The first half of the eighteenth century was a period of demographic stagnation, in which grain prices were relatively depressed and wage rates relatively high. As population growth took off after c.1760, however, arable prices rose, and they increased still further during the Napoleonic War period of 1793-1815. At the same time, the real costs of labour fell in the east of England, where the effects of demographic growth upon employment opportunities were accentuated by a decrease in the availability of rural by-employments, as the local textile industry went into steady decline in the face of competition from the industrialising north and west of England. With farm incomes rising and the costs of labour falling, it is hardly surprising that more and more farmers began to adopt underdrainage in the late eighteenth and early nineteenth centuries.

Outside the eastern counties, in the arable lands of the Midlands, the story was very different. Here underdrainage spread much more slowly. William Marshall in 1796 implies that in the Leicestershire/Warwickshire clays it first appeared in the 1760s, and the same seems to have been true in the clay vales of Oxfordshire and Buckinghamshire (Marshall, 1796, pp. 141-20; Wordie, 1985, p. 341). Even then, it was only sporadically employed: underdrains were difficult to install where farmland still lay in intermingled open-field strips, as was still often the case in the Midlands. In such circumstances drainage usually continued to be by ridging, although in the nineteenth century underdrains sometimes replaced the open furrows flanking the lands (Kain and Meade, 1977, pp. 135-6). Before the great wave of Parliamentary enclosure in the 1770s and 1780s, moreover, enclosure of heavier land had usually been associated with conversion to pasture. Parliamentary enclosure in this region also often led to the laying down of ploughland to pasture. But not always: and where arable land use continued, bush drains were sometimes installed in the newly-enclosed fields. Allen suggests that this improvement was one of the main factors behind the increase in yields recorded in a number of south Midlands parishes following enclosure (Allen, 1994, p. 120). Underdrainage remained, however, essentially a practice of Essex, Norfolk, Suffolk and the adjacent counties, until the spread of tile drainage in the middle decades of the nineteenth century.

– THE END OF BUSH DRAINAGE –

Across much of arable England, especially in Essex, Hertfordshire, and East Anglia, underdrainage was thus widespread by the middle decades of the nineteenth century. Bush drainage was an efficient and workable method which was only gradually replaced by ceramic pipes. It is true that earthenware tiles and pipes were coming into use early in the nineteenth century. Arthur Biddel of Playford in Suffolk purchased 'one thousand draining tiles from Goodings of Tuddenham' in December 1817 (Thirsk and Imray, 1958, p. 40). But only in the 1850s, following improvements in manufacture and the start of government loan schemes, did they come into widespread use. As late as 1852 Caird implied that tiles had as yet made little headway on the Suffolk clays, although pipes were generally used for the main drains, and that 'in all cases where it is found desirable that the work should be permanent, pipes or tiles are used throughout' (Caird, 1852, pp. 152-3). Some estates in Norfolk and Suffolk seem to have shown remarkably little interest in the improvement. The first time draining appears in the records of the Flixton estate in Suffolk, for example, was as late as 1862, when £17 was spent at Fressingfield and £123 at Flixton Grange on underdraining, ditching, and guano (ESRO, HA12 D/1/9/9). When John Keary prepared proposals for the Duke of Norfolk's estates in south Norfolk in 1861, one of the main deficiencies he highlighted was the relative absence of tile drainage, noting that the tenants still relied on the traditional, cheaper methods (NRO Smiths Gore 20.10.70 1520). Clearly, on the very eve of the agricultural depression, some large clayland estates were taking only moderate interest in the innovation, their tenants continuing to drain by the older method. Indeed, bush drains continued to be employed on some Norfolk and Suffolk farms up until the time of the Second World War: the method was still being recommended by the Ministry of Agriculture as late as 1925 (Evans, 1956, p. 120; Evans, 1970, p. 87; *Journal of the Ministry of Agriculture*, 1925, pp. 986-8).

Nevertheless, by the second half of the nineteenth century tile pipes had come into widespread use throughout the East, and the older method was, for the most part, restricted to the smaller farms, and smaller estates, lacking the capital necessary to implement more permanent schemes.

In this relatively limited area of England the adoption of tile-pipe drainage in the middle decades of the nineteenth century thus represented an improvement of an existing practice rather than a radical new departure. It was in other regions, barely touched by the development of bush drainage, that tile drainage made its greatest impact.

— THE IMPACT OF UNDERDRAINAGE —

As is well known, the 'agricultural revolution' of the late eighteenth and early nineteenth centuries involved two main processes: an expansion in the area of land under cultivation, and an increase in yields per acre. Less widely recognised is the fact that this period also saw an important transformation of England's agrarian geography. In the sixteenth and seventeenth century large areas of heavy clay soils in eastern England – in the old-enclosed counties of Essex, Hertfordshire, Norfolk and Suffolk – had been devoted to pasture farming – to dairying and to bullock rearing – and in many districts farms had no more than 25 per cent of their land in tilth. In the second half of the eighteenth century, however, the area of arable steadily expanded. Pasture closes fell to the plough, not only on the lighter clays but also on the heaviest soils, some areas of which had perhaps never previously been cultivated. By the 1840s, on many farms, the proportion of arable and pasture had been more or less reversed.

Bush drainage played a vital part in this transformation. Much of this new arable land could simply not have been successfully cultivated without underdrainage. But the technique had another, and arguably more important, role. New rotations were increasingly implemented in the course of the eighteenth century, in which cereals were alternated with courses of clover or 'artificial' grasses, and turnips. The clover fixed nitrogen directly from the atmosphere; both crops provided an increased supply of fodder which allowed livestock numbers, and thus supplies of manure, to be maintained (although probably not significantly increased) even though the area under pasture steadily declined. Indeed, without the adoption of these new crops it is doubtful whether the area of arable could have been expanded, without a decline in the numbers of livestock kept, and thus in the amount of manure produced. A fall in yields per acre would have occurred at the same time as the area of cultivation expanded.

The new rotations were thus essential to the expansion of cultivation on the claylands. But they could not easily be implemented on waterlogged soil. Turnips are more susceptible to disease in damp soils and even if grown successfully cannot readily be either 'pulled', and taken to yards; or directly fed off in the fields; because of the tenacity of the soil and the effects of poaching. In many contexts, therefore, the new rotations could not have been adopted without improvements to field drainage.

There is little doubt that as well as allowing the large-scale adoption of new crops, and an expansion in the area under cultivation, the widespread adoption of bush drains also served to raise cereal yields in its own right. Between 1750 and 1850 there was a steady increase in yields per acre for both wheat and barley on the heavier soils of the eastern counties, and while many innovations contributed to this, contemporaries certainly believed that improved drainage was a major factor. Glyde, for example, suggested that underdrainage had increased wheat yields on the East Anglian claylands by 33 per cent – from 24 bushels per acre to 32 – between 1750 and 1850 (Glyde, 1856, p. 338). Bush drainage is thus, arguably, one of the most neglected innovations in English farming in the 'agricultural revolution' period, and while the practice had some impact in other areas of England it was on the heavy boulder clay soils of Essex, Hertfordshire, Norfolk and Suffolk that its effects were most profound, rendering them the one of the most important corn-growing districts in England.

CHAPTER 5

Arable land drainage in the nineteenth century

A. D. M. PHILLIPS

During the nineteenth century, underdraining was subject to dramatic revaluation as an agricultural improvement. With the development around 1840 of new approaches to the layout of drains, and with the introduction of a fill that brought to drains a durability hitherto unknown, underdraining was promoted amongst agriculturalists as an effective and permanent improvement. However, the new systems of underdraining were credited not solely with the physical attribute of the successful removal of excess soil water but, more significantly as a consequence of drying land, with the potential for instigating economic change in heavy-land agriculture, especially in terms of cultivation costs, crop growth and yield, crop choice, and farming systems practised. For mid-century agricultural commentators, underdraining came to be regarded as the fundamental basis of agricultural advance on heavy land: as Andrew Thompson, a leading land agent, noted underdraining had become 'the first or foundation' of all farming improvement on such land (Phillips, 1975, p. 33). Many landowners and farmers came to share this view, and the capital input into underdraining between 1845 and 1899 has been estimated at about £27.5 million, a level of investment comparable to the outlay calculated by Holderness as having been spent on English parliamentary enclosure in public costs and subsequent expenditure on ditching, hedging and fencing (Holderness, 1971, pp. 165-7). The spread of the new systems of underdraining in the second half of the nineteenth century represented one of the major technical developments in English agriculture in the period 1700-1900.

— THE NEED FOR DRAINING —

To assess the impact of underdraining on nineteenth-century agriculture, a measure of the amount of land that would benefit from the improvement and its distribution is essential. While the dichotomy between heavy- and light-land farming was widely appreciated, few attempts were made by nineteenth-century agriculturalists to calculate the area in need of draining in England. Most contemporaries relied on a qualitative judgement of the extent of the improvement's need, exemplified by that of Josiah Parkes in 1845 that 'a most enormous and untold' quantity of land required to be drained (BPP, 1845, XVIII: q 138).

Those agricultural writers who added some quantification to such statements generally gave no indication of either the basis of calculation of the areas or their distribution; nevertheless, the results achieved currency in the contemporary literature. Thus, Philip Pusey reported in 1841 and 1842 that one-third of England, 10 million acres, required underdraining (Pusey, 1841, p. 103; Pusey, 1842, pp. 169-70). This figure corresponded to the acreage James Caird considered in need of draining in 1873, although he applied it to Scotland and Wales as well as to England (BPP, 1873, XVI: q 4126). Others suggested much larger areas: J. Bailey Denton in 1842 proposed that 10 million of the 12 million acres of arable in Great Britain should be drained (Denton, 1842, p. 64). This level of activity was adopted by Joshua Trimmer in 1847, who added that 15 million acres of pasture should be treated, creating a total of 25 million acres, about 75 per cent of the cultivated area of Great Britain (Trimmer, 1847, pp. 1-3). Although these high values may have emphasised the contemporary view of the need for the improvement, their range is a reflection of the lack of any thorough survey in their formulation.

The only detailed nineteenth-century analysis of the area requiring draining was that made by Denton, which appeared in 1855 and with modifications in 1883 (Denton, 1855, pp. 3-5; Denton, 1883, pp. 33-6). For individual and for groups of counties, he calculated the area of wet land which could 'be drained with advantage'. In 1855 he concluded that 15.3 million acres of England, 48 per cent of total area, needed draining. By 1883 Denton had enlarged the area of wet land to 16.5 million acres, or 50 per cent of the land area of England. The distribution of wet land was not uniform throughout the country, however, and at the county level at the latter date he saw Berkshire, Dorset, Hampshire, Oxfordshire and Wiltshire (in South-Central England); Norfolk and Suffolk (in the East); and Cornwall (in the South-West); as counties with the lowest proportions of land in need of draining. All the others had wet land acreages in excess of 40 per cent of total area, and in the case of Cheshire, Durham and Staffordshire over 70 per cent of total area.

Denton's estimate of wet land point to a large proportion of the cultivated area of England in need of draining: wet land in 1883 formed 69 per cent of the average cultivated area of the country in the decade 1870-9. However, an element of exaggeration entered Denton's calculations. His figures were derived from the solid geology of different parts of the country, and the wet-land areas represented the total acreage of geological formations with drainage difficulties. No distinction was drawn between cultivated and uncultivated land within these divisions. The wet-land acreage consequently incorporated significant areas that were not cultivated, either in 1855 or in 1883, and that as upland would not have warranted cultivation.

In the absence of other contemporary analyses of draining need, for present purposes use has been made of the Soil Survey's soil maps of England and Wales of 1975 and 1983 (Avery *et al.*, 1975; Soil Survey, 1983). These maps depict both the area and distribution of soils with drainage difficulties. Using the 1975 map, and removing areas of alluvium and lowland peat, of upland soils and peat, and of well-drained soils, the extent of clayey and loamy soils with impeded draining which require underdraining amounts by computation to 13 million acres, 41 per cent of the total area of the country, or 55 per cent of the cultivated area (Phillips, 1989, pp. 31-40). The density of these soils was highest on a county basis in the east Midlands, where they formed 65 per cent or more of the total area (Figure 5.1). Such soils were also significant in area in both the west Midlands and northern counties, with the exceptions of Derbyshire and Westmoreland. In general they formed a smaller proportion of the total area of eastern, south-eastern and southern counties, although relatively high levels were recorded for Suffolk, Middlesex, Surrey and Sussex. The south-western counties possessed the lowest quantities of clayey and loamy soils with impeded drainage. Although smaller than Denton's estimates, the area of soils with draining problems demonstrates the fundamental importance of underdraining to English agriculture in the nineteenth century.

— UNDERDRAINING ACTIVITY IN THE — EARLY NINETEENTH CENTURY

No precise evidence is available at the national level of the extent to which underdraining had been adopted on this area of wet land by the beginning of the nineteenth century, although some idea of the extent of activity around 1800 may be derived from the wealth of agricultural

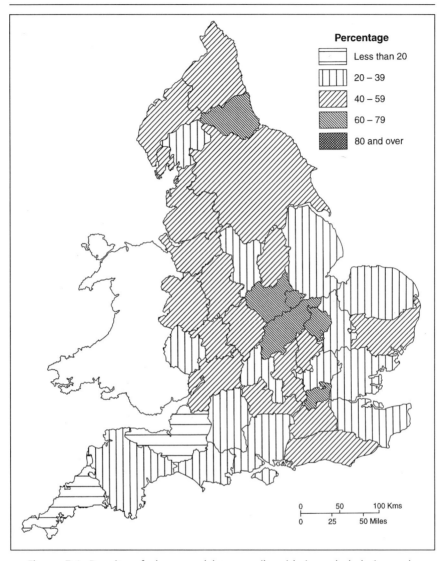

Figure 5.1 Density of clayey and loamy soils with impeded drainage by county. (Source: Phillips, 1989, pp. 36-9).

literature that appeared at the time, especially the two editions of *General Views* of the agriculture of each county sponsored by the Board of Agriculture (1793–1817), the *Annals of Agriculture* (1784–1815), the early numbers of the *Farmer's Magazine*, the *Communications to the Board of Agriculture* (1801–6), the report by the Board of Agriculture on *The Agricultural State of the Kingdom in . . . 1816* (1816), William

Arable land drainage in the nineteenth century

Marshall's *Rural Economies* of various districts of England (1787-98), and a number of draining treaties. While definite limitations must be recognised in the use of this evidence (Phillips, 1989, pp. 41-2), these literary sources would point to a broad distinction in underdraining activity between eastern counties and other parts of the country (Figure 5.2). The concentration of references in eastern England

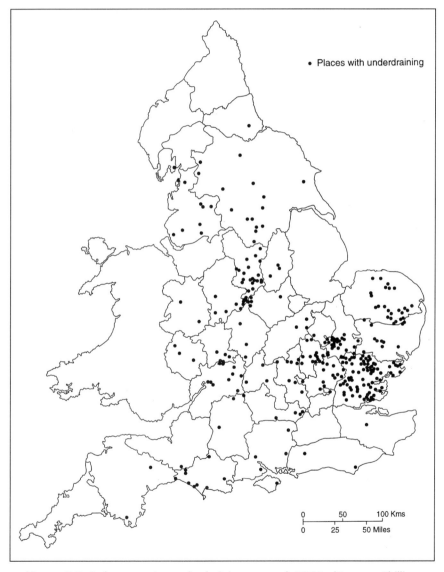

Figure 5.2 References to underdraining around 1800. (Source: Phillips, 1989, p. 44.)

would suggest that the improvement was widely adopted in that region, as described by Williamson in the previous chapter, while in the rest of the country its use would seem in general less intense, with greater attention being paid to Joseph Elkington's method of spring draining (Johnstone, 1801).

In the north of the country John Bailey and George Culley recorded that the improvement had only recently been introduced into Cumberland, Durham and Northumberland, and that in consequence little land had been drained (Bailey and Culley, 1805, p. 128; Bailey, 1810, p. 202). A similar level of activity was noted in Westmoreland and parts of Yorkshire. In the west Midlands, underdraining was again noted as having but recently spread. Although an element of variation in the intensity of underdraining was reported in the east Midlands, the general view was that the improvement had been little adopted. Arthur Young ascribed the absence of underdraining in Lincolnshire to its recent introduction (Young, 1799, p. 242). While 'the necessity of draining wet lands has of late years been much better understood and attended to', as Robert Lowe found in Nottinghamshire, 'the very great neglect' of the improvement, that Richard Parkinson observed particularly in Huntingdonshire, was widely noted (Lowe, 1798, p. 98; Parkinson, 1813, p. 292). These comments are very much at variance with R. C. Allen's claims based on the same sources of the widespread adoption of underdraining in the south Midlands by 1800 (Allen, 1992, pp. 117–21, 137–9). The lack of extensive underdraining was a theme common to all the reports on the south-western and south-eastern counties, the Revd Arthur Young observing that in Sussex, where the improvement was greatly needed, the methods were not thoroughly understood and the practice was confined to a few individuals (Young, 1808, pp. 191–7).

Throughout the East Anglian counties a rather higher level of underdraining activity was reported. Here, as explained in the previous chapter, the practice was noted to be widely understood and implemented. Arthur Young considered that in Norfolk the improvement was widely undertaken and in Suffolk the practice was 'general on all the wet lands of the county [and was] too well-known to need a particular description' (Young, 1804a, pp. 389–95; Wade Martins and Williamson, 1995, pp. 26–9). Similar levels of underdraining were noted in Cambridgeshire, Essex and Hertfordshire. In these counties, moreover, greater emphasis was given to underdraining as a means of removing surface water than in other parts of the country, and for that purpose underdrains were in general reported to be shallow, averaging between 20 and 30 inches in depth (50-75cm), filled with bushes, straw, thorns, wood and occasionally stones, while mole drains at about 20 inches deep were noted in all the counties. Elsewhere in the country, far greater attention was given to methods and examples of spring draining, as practised by Joseph Elkington, and the infrequency of references to underdrains of stone, turf and wood and to mole drains offers further support for the view that the large areas of heavy soils suffering from surface water had been little touched.

The general impression provided by contemporary agricultural literature was thus that while in East Anglia, and the adjacent counties, underdrainage was coming into widespread use by c.1800 overall little underdraining had been undertaken in the country at this time; and even in East Anglia, much land remained in need of the improvement.

One reason for this, according the contemporaries, was the responsibility for the implementation of the improvement. As already noted in the previous chapter, the systems of underdraining to remove surface water to be found in East Anglia possessed a limited life: their depths rendered them susceptible to disturbance in arable areas, while the dominantly vegetative materials used for fills were liable to easy blockage. This method of impermanent underdraining came to be regarded as part of the tenant's commitment to agriculture, and many commentators were of the opinion that this restricted both the spread and effectiveness of the improvement. Outside East Anglia few tenants would seem to have been prepared to accept the responsibility for adopting the improvement, and tenants resorted to alternative and simpler means of removing surface water from heavy lands, especially ridge and furrow, surface cuts and grips, and ditches. Despite limited efficiency and loss of cultivated land, extensive use of these systems was described throughout the country, and even in East Anglia Arthur Young found these techniques widespread on clay soils in Essex, Hertfordshire and Norfolk (Young, 1804a, pp. 110; Young, 1807, vol. 1, p. 16).

In the first half of the nineteenth century, while a growing perception of the need for underdraining can be identified in the agricultural literature, it is not accompanied by any evidence of a significant spread of the improvement. A number of witnesses before the Select Committees on agriculture in 1833 and 1836/7 spoke of the need for underdraining and described the introduction and some expansion of the improvement in individual Midland and southern counties during the 1830s. However, the rate of growth would seem to have been slow. The report of the 1833 Select Committee recorded that heavy clayland had lacked capital investment, while C. S. Lefevre, commenting on the evidence before the 1836 Committee, noted the relative absence of underdraining on such soils throughout the country (BPP, 1833, V, iv; Lefevre, 1836, pp. 13–14). Contributors to the first issues of the *Journal of the Royal Agricultural Society of England* wrote both of recent indifference towards the adoption of underdraining and, as we have already seen, of the extensive areas of cultivated land that would benefit from the improvement: sentiments reiterated in more detail by the authors of the early prize essays on the agriculture of individual counties also published in the *Journal*. Some support of this assessment may be provided by a recent analysis of landlord expenditure on underdraining over the nineteenth century on a sample of some 20 estates in Devon, Northamptonshire and Northumberland (Phillips, 1989, pp. 126–31) which reveals low per acre levels of provision up to the late 1830s (Figure 5.4a).

No single source provides statistical data that allow the extent of underdraining in England to be determined at the end of the 1830s. However, this lack may be partially remedied by recourse to local tithe agents' and assistant tithe commissioners' references to the improvement in their reports on tithe agreements arising from the Tithe Commutation Act, 1836, contained in the general body of tithe files (Kain, 1986, pp. 562–631). The availability of these reports is neither complete nor uniform over the country, their incidence being greatest in East Anglian counties, and in the western counties of Dorset, Herefordshire and Somerset, and least in Midland counties. A further problem is that discussion of the improvement was not mandatory, mention of underdraining in the reports depending on the enthusiasm and awareness of individual assistant tithe commissioners. The results from mapping those tithe districts where the practice was observed endorse the general trends indicated in the literary evidence, and from the sample estates, of the small-scale adoption of the improvement by around 1840 (Figure 5.3). Many tithe districts throughout the country were described as possessing land in need of underdraining, but underdraining with tiles or with more traditional materials, such as bushes, straw or stones, was recorded as being practised in much fewer tithe districts. Only Norfolk and Suffolk deviated from this pattern with numerous tithe district references to the improvement. The above-average underdraining activity reported in these counties at the beginning of the nineteenth century seemed to have remained unchanged by 1840. The system of underdraining employed had

Arable land drainage in the nineteenth century

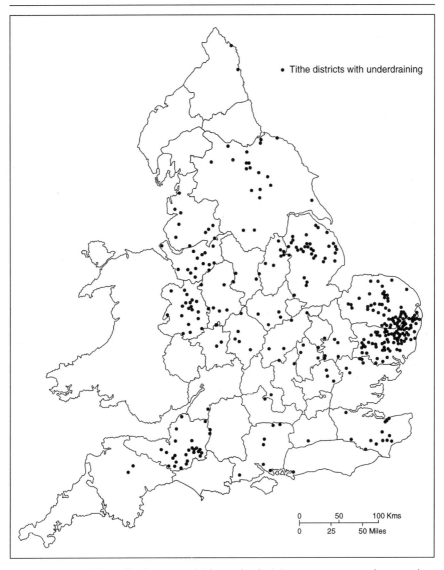

Figure 5.3 Tithe districts in which underdraining was reported around 1840. (Source: Kain, 1986, pp. 562-631).

altered little, and was still regarded as an element of a tenant's cultivation practice (Pusey, 1843, pp. 23-49). Yet (*pace* Williamson, see Chapter 4) it remains questionable whether these frequent references to the improvement means that that very extensive areas of these counties had been underdrained by 1840. Although contemporaries considered the

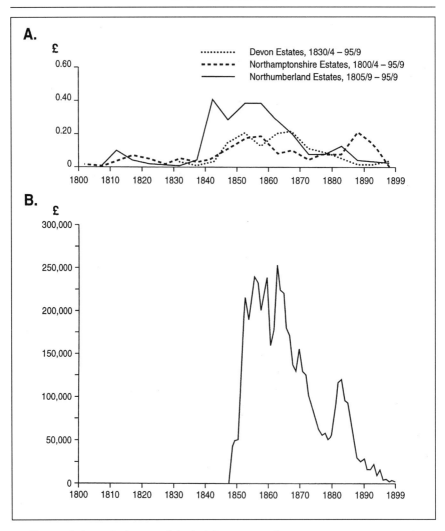

Figure 5.4 a. Quinquennial draining expenditure per acre on sample estates in Devon, Northamptonshire and Northumberland, 1800/4 to 1895/9. b. The supply of loan capital for underdraining in England, 1847–99. (Source: Phillips, 1989, pp. 123–30.)

extent of tenant involvement in underdraining in these counties exceptional, much of the clayland could still be described by agents working in this area as in need of the improvement at the middle of the century. And accounts of the level of landlord capital provision for underdraining were more in accord with conditions found in other parts of the country (Wade Martins, 1980, p. 96; Caird, 1852, pp. 133, 145).

— REVALUATION OF THE IMPROVEMENT —

Against this background of need, but limited implementation, from the late 1830s and 1840s there occurred a rapid spread of underdraining in response to changes in the technology, management and financing of the improvement. Initially a revaluation of the improvement was described in the agricultural literature of the time, involving a reappraisal of underdraining systems, materials, costs, permanence and productivity. During the 1820s and much of the 1830s, specific draining and general agricultural accounts reiterated the view prevalent at the beginning of the century, that wetness in land was capable of being treated by a variety of draining systems depending on the origin – from springs or rainfall – of that water. From 1831, and particularly from 1836, in evidence before the Select Committee into the state of agriculture, James Smith presented a systematic and unified approach to the problem of excess soil water (Smith, 1831, pp. 4-15; BPP, 1836, VII, qq 14976-15233). For the first time he identified the removal of surface water as the prime objective of the improvement, but his system of thorough draining, which comprised throughout a field a series of frequent, parallel drains at a minimum depth of 2½ft (76cm) and at intervals dependent on soil texture, could also deal with any spring water that might arise. Modification of this system was suggested from the early 1840s, chiefly by Josiah Parkes, who argued that drain depths should be increased to 4ft (1.2m) (Parkes, 1846, pp. 249-72). Such additional depth not only allowed the laying of drains at greater intervals, so reducing labour and material costs, but was also considered to achieve a more effective draining of soil water. Although there was debate over the merits of respective depths, a coherent system of underdraining was recognised that was applicable to most conditions of soil wetness.

Associated with this new approach, the literature disclosed the beginnings of the mechanisation of the production of the most reliable of drain fills, a process reported to reduce total costs and to increase permanence. By the mid-1830s a range of drain types was known. Mole draining was regarded as the cheapest, being priced at 18s per acre in 1821. Bush and straw drains varied between £1 10s and £2 10s per acre over the 1820s and 1830s in East Anglia, while plug and turf drains ranged between £1.10s and £2.10s per acre. Stone and tile drains were more costly although they lasted longer. With stones readily available and drains laid at 20-ft (6-m) intervals, stone draining was estimated at £6 per acre under Smith's system. Semi-circular tiles laid on soles were recognised as a more effective form of drain but were expensive. In 1841 tiles and soles were reported as costing between £2 and £3 per 1000 in Berkshire, Gloucestershire and Norfolk. As 2000 tiles would be required to drain an acre at 20-foot intervals, tile draining would require an outlay of £4 and over for materials alone (Phillips, 1989, p. 159).

The high price of tiles stemmed from their being moulded and made by hand. With the invention of machines that mass-produced tiles and soles, costs were much reduced. In 1835, Robert Beart of Godmanchester, Huntingdonshire, introduced such a machine, to be followed by several others. On average these machines were claimed to produce tiles and soles from £1 to £1 5s per 1000, lowering significantly the cost of tile drains (Beart, 1841, pp. 93-9). The use of tiles and soles as fill was superseded by the development of cylindrical drain pipes. Handmade drain pipes had been reported from various parts of the country from early in the nineteenth century. The existence of several simple machines that produced limited numbers of drain pipes was noted by 1840 in Suffolk, Essex, Kent and Sussex. The design of these was improved to provide greater output and standardisation of form. The first of these new drain-pipe machines was credited to Thomas Scragg of Calveley, Cheshire in 1842, but the growth in their number was swift (Parkes, 1843, pp. 369-79; Parkes, 1845b, pp. 303-23).

While the drain pipe was acknowledged to represent a technical advance over the tile and sole, in that the likelihood of blockage was reduced, the new machines were regarded as contributing to lowering the overall cost of underdraining. In 1845, pipes with a diameter of 1 in (2.5 cm) were priced at 12s per 1000. At this price both Josiah Parkes and Philip Pusey could claim that heavy clayland could be drained using the new systems for no more than £3 per acre (Parkes, 1845a: 125-9). Although doubts were raised as to the ability of 1-in pipes to maintain an open channel underground – 2-in (5-cm) pipes, which were priced at £1 to £1 5s per 1000, and thus increased average costs to around £5 per acre being considered safer – effective underdraining was reliably reported as being obtainable at the same rate as traditional methods.

Added to these cost reductions, the new underdraining systems were attributed with a high degree of permanence. Although views of their duration varied, mole, bush, straw, turf and plug drains were acknowledged as temporary improvements with the need for regular renewal. However, many agriculturalists, impressed by the technical efficiency with which pipes were produced, saw the new systems as possessing a much longer, if not perpetual, life. While recognising a finite existence to the improvement, the 50 years suggested by Josiah Parkes represented a life expectancy at least twice that of traditional draining methods (Pusey, 1842, pp. 169-172; BPP, 1845, XVII, qq 561-3).

Technical efficiency, permanence and low cost of drains were linked in the agricultural literature to increased productivity, which was largely assessed in terms of crop yields, especially those of wheat. Although a considerable number of estimates were made, landowners were led to believe that a growth in wheat output of between 10 and 30 per cent per acre could be reasonably expected on the adoption of these new draining methods. Such productivity was further seen as providing landowners with an opportunity of obtaining a clear return on the capital investment in the improvement in the form of increased rent. Estimates of this return varied in the 1840s but they were rarely less than 10 per cent and often as high as 25 per cent. In sum, the change in draining systems that occurred in the late 1830s and 1840s was presented not simply as a technical advance resulting in the more effective removal of excess water in soils, but also as an economic improvement that because of increases in productivity had real implications for enhancing land values (Phillips, 1975, pp. 253-4). Whether or not the claims made for the new underdraining systems were justified, the fact that such views were expressed throughout the agricultural press by the leading agricultural authorities of the day must have raised contemporary perceptions of the value of the improvement.

At the same time, these technical developments necessitated a change in the management of the improvement. The increase in costs, the need for a degree of care and supervision for effective implementation, the predicted permanence, and the potential for enhancing rents made the improvement ideally suited to landlord finance and control. Yet many contemporaries perceived that not all landowners, especially those of settled estates, were able to provide the capital to undertake the improvement, and during the middle decades of the nineteenth century, legislation was introduced to facilitate landlord investment in underdraining (Spring, 1963, pp. 143-61).

In 1840 and 1845, acts were passed which established the principle that tenants for life might borrow for underdraining on the security of their estates, repaying capital and interest in a rent charge over a period of years. This principle was developed more effectively in the Public Money Draining Act of 1846. Introduced by Sir Robert Peel in his repeal speeches as a scheme both to increase agricultural output directly and to promote a general spirit of agricultural improvement so as to offset the price falls that might result from the loss of the Corn Laws, the Act made available £2 million from the Treasury for landowners of fee simple as well as settled estates to borrow for the purpose of underdraining. In 1850 a further £2 million was sanctioned for underdraining. Based on the same principle, the Private Money Draining Act of 1849 was introduced to allow landowners to borrow from private sources, as distinct from the government, in order to undertake underdraining. Repealed in 1864, its terms were incorporated into the Improvement of Land Act of 1864. In addition to governmental and private sources of loan capital, five land improvement companies were established by acts between 1847 and 1860 to apply collective capital, largely from insurance companies, to agricultural improvement on landed estates. While not all landowners made use of monies from the land improvement acts to implement underdraining on their properties, the existence of the legislation in itself did much to encourage landlord provision of the improvement, and between the date of the first loan in 1847 and 1899, £5.499 million had been borrowed for the purpose of underdraining in England (Phillips, 1989, pp. 50-62, 73-81).

—THE ADOPTION OF UNDERDRAINING IN THE— SECOND HALF OF THE NINETEENTH CENTURY

By the late 1840s underdraining had emerged as the outstanding agricultural improvement of the day. New technologies of draining had made the effective, permanent removal of excess soil water a reality and had brought the promise of enhanced agricultural productivity; they had transformed the improvement into a predominantly landlord responsibility; and they had established state support for landlords in its financing. The precise extent to which landlords and tenants responded to the opportunity of improving the wet lands of England is impossible to determine in the absence of any comprehensive record of capital outlay on underdraining by the large body of landowners in the second half of the nineteenth century. However, records of the loans for underdraining under the land improvement legislation, amounting to £5.499 million, have in the main survived. Although they do not represent all underdraining activity, a great many landowners preferring to finance the improvement from their own resources, they provide a coherent source for the analysis of the adoption of underdraining over the period. Contemporary agriculturalists argued that the pattern of loan-financed underdraining was representative of all underdraining activity in the country, a view endorsed by a recent study of the improvement (BPP, 1873, XVI, qq 586-9, 4125-30; Phillips, 1989, pp. 117-19). As such, records of loans may be used to construct at the national level an index of the rate, amount and distribution of land drained between 1847 and 1899.

The supply of loan capital for underdraining under the land improvement acts was not uniform, and distinct periods of investment may be identified (Figure 5.4b see p. 62). The two decades 1850-69, with 68 per cent of all loan capital for draining, witnessed the most rapid spread of the improvement. The implementation of underdraining funded by these loans was supervised by the Inclosure Commissioners. Convinced of the efficacy of Parkes' deep draining system and of pipes for fill, although insisting on the use of those with diameters of 2in (5cm) and greater, they applied the method rigorously up to the late 1870s. Many estates not employing loan capital also adopted the system favoured by the Inclosure Commissioners (Phillips, 1989, pp. 206-9). Although a few examples of draining failure were reported, the reliability of the draining systems laid down by the Inclosure Commissioners was high. Leaving aside the implications for agricultural productivity, the concentration of highly effective underdraining schemes between 1850 and 1869, producing faster and more complete run-off from the land and therefore greater discharge of water, must have created new and extensive problems for the management of rivers and other arterial channels in areas of greatest activity (Denton, 1863).

After 1870, the supply of loan capital for draining began to fall, interrupted by renewed investment in the 1880s, with 13 per cent of all loan capital for draining (Phillips, 1989, p. 124). From the late 1870s the Inclosure Commissioners began to relax their insistence on deep draining, and sanctioned drains at shallower depths (not less than 3ft) (0.9m) but at closer intervals, a modification also found on many estates funding the improvement privately. The adjustment reflected changing attitudes to the purpose of draining. A number of new draining treaties emphasised the importance of concentrating on the removal of surface water alone, which could be more effectively achieved by shallower, closer drains (French, 1879, pp. 165-7; Phillips, 1989, pp. 211-14). Moreover, a series of wet seasons in the late 1870s and early 1880s, which encouraged a renewed interest in the improvement, must have convinced not only landlords and tenants but also the Inclosure Commissioners of the need to adopt this policy.

In the last decade of the nineteenth century, less than 2 per cent of all loan capital for draining was employed, suggesting a general falling off in the implementation of the improvement in the country. Some demonstration that overall this time series reflected general trends in draining activity in England in the second half of the nineteenth century may be given by examining landlord per acre expenditure on the improvement on the sample of estates in Devon, Northamptonshire and Northumberland (Figure 5.4a). For all three groups of estates, landlord outlay was greatest between 1840 and 1869. After 1870 investment fell markedly, with some increase in expenditure in the 1880s.

The loan capital for draining may also be used to estimate the amount and distribution of land drained in the second half of the nineteenth century. A recent analysis of underdraining in England has suggested that loan capital commonly constituted about 20 per cent of total capital outlay on the improvement, the remainder being supplied by landlords from estate revenues. Assuming that this proportion is applicable throughout England, it is possible to provide an estimate of total investment in underdraining at both the national and county level. Applying this ratio to the £5.499 million of loan capital for draining provided under the land improvement legislation in England, it may be suggested that underdraining absorbed £27.5 million of landlord capital between 1847 and 1899 (Phillips, 1989, pp. 101, 107, 116-21).

Conversion of this sum into acreage equivalents is difficult because the cost of draining varied both in space and time over the period. Contemporary estimates ranged from £4 to £8 per acre, but perhaps the most reliable large-scale guide to average draining costs may be found in the surviving reports of Andrew Thompson on draining carried out on 133 estates with loan capital. On these properties draining averaged £6 per acre (Phillips, 1975, pp. 260-3). At this rate, some 4.583 million acres (1.856 million hectres) are estimated to have been drained in the second half of the nineteenth century. This conjectural figure would have resulted in 35 per cent of the land in need of draining being treated. While the scale of investment in the period on the wet and heavy lands of the country is impressive, nevertheless only a part of the land in need of the improvement benefited. The farming advantages associated with the new systems of underdraining were not to be experienced on all the wet lands of the country.

If not universal, the loan capital data points to great spatial diversity in the adoption of the improvement (Figure 5.5). Capital investment in underdraining would seem to have made least progress in eastern, and to a lesser extent south-eastern and extreme south-western England, but would seem most developed in northern, west Midland and western counties. The draining loan data would indicate that less than 15 per cent of soils in need of draining in Essex, Norfolk and

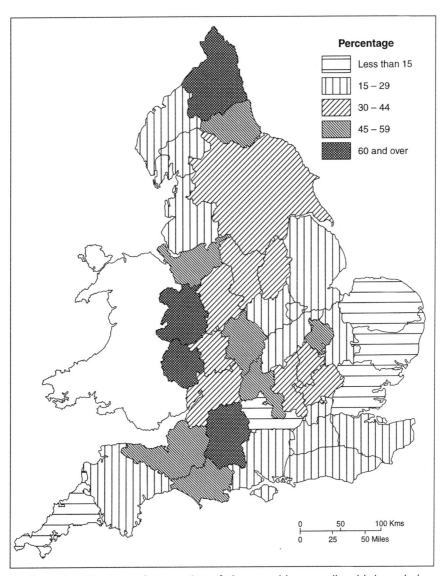

Figure 5.5 Conjectural proportion of clayey and loamy soils with impeded draining underdrained in the second half of the nineteenth century. (Source: Phillips, 1989, p. 119.)

Suffolk were treated between 1847 and 1899, compared with over 50 per cent in Durham and Northumberland, and in Herefordshire, Shropshire and Worcestershire. In general, underdraining activity was most intense in those parts of the country, the North and West, that in the third quarter of the nineteenth century were marked by the highest per acre incomes for landowners and farmers, and experienced according to James Caird the highest land values (Caird, 1880, pp. 98-9; Grigg, 1965, pp. 62-3; Thompson, 1981, pp. 106-7). The adoption of the improvement may be seen not only as a product but also as an element of the prosperity of these areas. The apparent failure of underdraining to be widely adopted in East Anglia demands comment, especially as soils with impeded drainage formed a significant proportion of the region and as the area had frequently been noted in the literature up to the 1840s for the tenant systems of underdraining practised there. However, the suggested low level of landlord investment in underdraining is consistent with Holderness' identification of a general slackening of landlord involvement in agricultural improvement in Norfolk and Suffolk from the 1830s (Holderness, 1972, pp. 439-41). If underdraining were more widely spread in East Anglia in the second half of the nineteenth century than proposed here, it must have taken place outside landlord supervision and control and through the continuation of the tenant system with all its drawbacks for efficiency, durability and capital availability (Read, 1858, p. 270).

Although detailed explanation of these temporal and spatial patterns is beyond the scope of the present survey, some discussion of the forces acting on the adoption of the improvement in the period is possible. As underdraining after 1840 was promoted as a landlord improvement, it was most extensively and successfully developed on estates that recognised and accepted the financial and administrative implications of the new draining systems. Not all estates identified the need or were willing to take responsibility for the improvement. Some estates invoked a tenant contribution in the form of either an equal capital sum or provision of labour to undertake the improvement. On these estates both the amount and efficiency of underdraining were lower than on those where the improvement was a landlord charge (Phillips, 1989, pp. 167-73).

While most effective as a landlord improvement, the intensity and occurrence of underdraining were limited by the capital sums estates were prepared to provide. The draining loan data indicate that capital supply for the improvement was strongly related to estate size. Underdraining was most readily and widely adopted on large estates, which overall possessed greater financial resources for the support of agricultural improvement. The provision of draining capital would seem to be generally ignored on smaller properties, especially those under 1,000 acres (405ha), where rental incomes offered less scope for extensive agricultural investment. On these, which formed a considerable part of the cultivated area of the country, waterlogged land must have remained in a largely unimproved state. Estate size emerges as a significant element in understanding both the spatial and temporal diffusion of landlord-financed underdraining (Phillips, 1989, pp. 173-85).

Landlord involvement in draining was generally confined to the supervision and capital supply of the improvement. In the main period of draining activity, 1840-70, as capital was normally only provided for the improvement when tenants agreed to pay interest on the outlay, the tenants must be seen as the active regulators of the amount and type of land drained. Tenants were forced to assess the adoption of underdraining largely in commercial terms. The improvement presented a means to increase crop output and income from waterlogged land, but at the same time necessitated the payment of interest. Only land that when drained could produce – through increased yields – a return that would cover their financial commitments to the improvement, and also more importantly provide a profit, would be attractive to tenants. As a result, the improvement was concentrated on good quality agricultural land, which when cleared of excess water had a high natural fertility; and on arable, which on draining afforded a greater and more immediate return than grassland. Of the arable crops, wheat offered up to the 1870s the largest return on interest payment. As the supply of suitable land declined, and as arable prices began to fall after 1880, draining became a less viable commercial proposition for tenants and demand for the improvement dwindled. Such underdraining undertaken towards the end of the century tended to be landlord sponsored, being largely remedial action to preserve land capability and quality, and thereby tenants (Phillips, 1989, pp. 193-205).

— CONCLUSION —

Nineteenth-century agriculturalists were keenly aware both of the extent of wet, heavy land and of the problems of its cultivation. They recognised the role of underdraining as the process that could transform agricultural practice on such land. Yet, despite its acknowledged importance, the effective underdraining of wetland areas in the country was far from complete at the end of the century. Before the 1840s, adoption of the improvement would seem restricted by the lack of effective systems of underdraining and reliance on tenant capital for its implementation. The perfecting of techniques of underdraining in the 1840s, together with the development of landlord responsibility for its management, was associated with widespread adoption of the improvement. However, the precise pattern and extent of adoption were dependent on the landlords' willingness to invest in the improvement and tenants' perception of its value. Landlord and tenant enthusiasm for the improvement was far from uniform throughout the country, and in the last two decades of the century began to dwindle. It is estimated that from the 1840s about 35 per cent of the wetland area of the country was effectively underdrained, with intensity being highest in parts of northern and western England. The wetlands of these areas would have enjoyed the benefits that underdraining could bring to agricultural activity. In other parts of the country, especially much of south-eastern and south-western England, the problems of wetland cultivation must have still been widespread at the end of the nineteenth century.

Chapter 6

Wetland soils

David Dent

— Distinctive processes —

Continual waterlogging determines the lifestyle of wetland communities. On the one hand there is plenty of water; on the other, waterlogging means a shortage of oxygen for aerobic respiration – the oxidation of organic substrates to release the energy needed for life. Wetland plants are specifically adapted to transfer air to their root systems, but the micro-organisms that decompose this organic matter and recycle the nutrients taken up by plants and animals do not have this facility. Instead they live anaerobically (Table 6.1), reducing dissolved nitrates to ammonia and nitrogen gas, sulphates to hydrogen sulphide and carbon dioxide to marsh gas – so wetlands bubble and stink. Will-o'-the-wisp is caused by the spontaneous ignition of marsh gas in contact with the air.

Anaerobic bacteria also reduce iron oxides, which colour well-drained soils red and brown, to soluble colourless iron (Fe^{2+}) ions. So wetland soils are drab in colour. The reduced iron may travel distances of several kilometres in solution, to deposit again as ochre or bog iron at the edge of the marsh or bog, or in drains and ditches where oxidising conditions prevail. Mottles and nodules of red/brown iron oxides and black manganese oxide within a drab soil matrix indicate a soil that has experienced alternating reducing and oxidising conditions. Rusty mottling is often conspicuous around burrows and old root channels that allow air or oxygenated water to enter the soil. In Britain, soils tend to be wetter in winter than in summer and a mottled soil layer indicates the zone of fluctuating watertable. Soils that are flushed by brackish or salt waters, which are rich in sulphates, accumulate intensely black iron monosulphide and, ultimately, pyrite and other metal sulphides.

Reduction processes consume acidity (Table 6.1), so the reaction of waterlogged soils is often near neutral. Even more significantly for plant growth, one key nutrient, phosphate, is much more soluble in reduced soils. Many wetlands also receive other nutrients in solution from the adjacent uplands so, with water also abundant, plant growth can be lush. In general, anaerobic decomposition of the plant matter thus produced is less effective than aerobic decomposition, so wetlands accumulate organic matter and, in freshwater conditions, extensive deposits of peat are able to accumulate.

TABLE 6.1 Some reduction-oxidation reactions in soil, in sequence of the energy released by reaction and order of reduction

The aerobic path in well drained soils

O_2 + $4H^+$ aq + $4e^-$ → $2H_2O$ l
oxygen   acid   electrons

Anaerobic paths in waterlogged soils

$2NO_3^-$ aq + $12H^+$ aq + $10e^-$ → N_2 g + H_2O l
nitrate   nitrogen

MnO_2 s + $4H^+$ aq + $2e^-$ = Mn^{2+} aq + $2H_2O$ l
manganese dioxide

$FeO.OH$ s + $3H^+$ aq + e^- = Fe^{2+} aq + $2H_2O$ l
iron oxide

SO_4^{2-} aq + $10H^+$ aq + $8e^-$ → H_2S g + $4H_2O$ l
sulphate   hydrogen sulphide

CO_2 aq + $8H^+$ aq + $8e^-$ → CH_4 g + $2H_2O$ l
carbon dioxide   marsh gas

Respiration involves a flow of electrons from the substrate, usually organic matter, to an electron sink. In well drained soils, the preferred sink is oxygen, which is reduced, combining with hydrogen ions to form water. In waterlogged soils, a range of alternative electron sinks may be used.

Management of temperate wetlands, such as those reclaimed in Britain, has usually involved drainage to make the soil suitable for pasture or arable, and this releases the nutrients stored in the organic matter and sediment. In their more natural state, however, wetlands also perform valuable functions: retaining floodwater and sediments, removing nitrate, retaining other nutrients and also pollutants, and the value of these functions is increasingly being recognised.

— SEDIMENTARY PATTERNS —

Wetland soils exhibit no less variety than the ecosystems to which they are bound; both reflect the hydrology, topography and, importantly, the history of sedimentation and erosion.

Coastal plains and estuaries experienced a rapid rise in sea level of some 30 m through global warming and ice melt, between 10,000 and 5,000 years before present. The rate of sea level rise has gradually declined although with periodic fluctuations. However, the North Sea Basin is itself subsiding so the current rate of sea level rise here is still between 1 and 2 mm/year (Shennan, 1989). In contrast, northern Britain is still rebounding from the weight of ice it bore during the Ice Age, resulting in raised beaches and rivers downcutting through valley terraces.

Over the last 4,000 or 5,000 years, wetlands adjacent to the North Sea and Irish Sea have experienced periods of relatively stable sea level alternating with periods of rising sea level accompanied, respectively, by the formation and breaching of coastal sand and shingle barriers. Extensive 'fen peat' accumulated in the sheltered freshwater lagoons and in some areas peat accumulation gradually raised the land surface above the regional groundwater level. These 'raised bogs' depend mainly on rainwater, so have much lower nutrient status than the surrounding fens.

Breaching of coastal barriers led to renewed tidal flooding, sometimes erosion of the peat, sometimes burial beneath mineral alluvium carried in by the tides. Active tidal creeks build up 'levees' where the incoming muddy water spills over their banks. The coarsest, often shelly, material is deposited close to the channel as the speed of the water is slowed. The finest clay gradually settles out in the intervening backswamp basins and, as sea level stabilises once again, the land surface is built up towards high-tide level and the coastline progrades.

A similar pattern of levees and backswamps builds up in river flood plains. All alluvial soils exhibit characteristic sedimentary layers, coarse-textured layers being deposited from swiftly moving waters, and clays settling out from still water. Organic-rich layers in the alluvium represent periods of stability when peat or an organic-rich topsoil built up, only to be buried by later sediments. Such patterns are well illustrated by the 1:250,000 scale soil maps of the Soil Survey of England

and Wales (1984). One of the best examples of the levee and backswamp pattern is shown by the 1:25,000 soil map of the Chatteris District of the Cambridgeshire Fens (Seale, 1975). Surveys of lowland peat in England and Wales are recorded by Burton and Hodgson (1987).

— SOIL RIPENING —

Clay and organic materials that accumulate under water have a very high water content, more than 75 per cent by mass and sometimes up to 95 per cent. Individual platy clay particles arrange themselves in an open, cellular structure which is easily deformed, giving the soil a buttery consistency. As sedimentation raises the land surface above the water level for longer periods, and as colonising vegetation extracts water from the soft soil, the cellular structure collapses and individual clay plates are rearranged face-to-face to give a denser, firmer soil of progressively lower water content. This process is known as 'soil ripening'. In natural wetlands, ripening progresses furthest on the levees (the highest, driest parts of the landscape), which eventually build up a brown nearly ripe topsoil, and least in the backswamps, which remain only half ripe.

All soil profiles continue to grow from the bottom upwards as fresh material accumulates at the soil surface. Tables 6.2, 6.3 and 6.4 illustrate the range of soil profiles found in natural wetlands.

TABLE 6.2. Soil profiles of natural wetlands: saltmarsh

Classification	Unripe saline clay (Dent, 1986)
	Calcaric Fluvisol (FAO/ISRIC, 1998)
	Friskney Series
Location	North Wooton, Norfolk
	c.80m west of 1966 embankment
	TF 610 275
Vegetation	*Puccinellia maritima* and *Aster tripolium*. On levees *Halimione portulacoides, Suaeda maritima, Cochlearia anglica*. In pools *Spartina anglica* and *Salicornia europaea*
Parent material	Marine alluvium. Laminated clay, silt and fine sand overlying fine sand
Drainage	Flooded twice a day during spring tides, not flooded during neap tides

Wetland soils

Brief description	70cm silty clay with thin layers of fine sand overlying loamy sand. Upper 30cm nearly ripe, brown, with abundant roots; below very dark grey, progressively less ripe with black mottles. Common shell fragments (5–10% $CaCO_3$) and saline throughout
Profile description	
Go (Cg1) 0–30cm	Brown (2.5YR 3/2) clay loam; nearly ripe, moderate coarse crumb; many fine roots; many remains of *Puccinellia* and few stems of *Spartina*; at depth some rust mottles; EC 6mS cm^{-1}, pH 7.0; gradual boundary
Gro (Cg2) 30–50cm	Very dark grey (2.5YR 4/2) clay loam with thin layers of sand; half ripe; very coarse prismatic and weak angular blocky; many remains of *Spartina* many roots of *Puccinellia*; at 30–40 cm yellowish red (5YR 6/8) iron oxide pipes around some vertical channels; channels of *Corophium* and *Nereis*; EC 20 mS cm^{-1}, pH 7.2; gradual boundary
Gr (Cg3) 50–70cm	Very dark grey (N3) inside peds and greyish brown (2.5YR 5/2) on ped faces; loam; half ripe; moderate coarse prismatic; dark red (5YR 3/6) coatings on ped faces; common fine roots, stems of *Spartina*; EC 20 mS cm^{-1}, pH 7.2;
2Gr (Cg4) 70+cm	Dark grey (N3) sandy loam with grey (5YR 5/1) along channels; common fine roots, live *Heteromastus filiformis* at 70cm depth; EC 20 mS cm^{-1}, pH 8.0

Note: Horizon nomenclature after Dent (1986) FAO alternatives in parentheses

TABLE 6.3 Soil profiles of natural wetlands: reedswamp

Classification	Unripe sulphidic muck (Dent, 1986) Thionic fluvisol (FAO/ISRIC, 1998)
Location	Washland on the River Waveney, Black Mill Pump TM 478 959
Vegetation	Reed swamp, dominantly *Phragmitis australis*
Parent material	Brackish water mineral alluvium
Drainage	Very poorly drained, tidal

Salts	Slightly saline throughout
Brief description	Dark grey, practically unripe silty clay over sulphidic half ripe silty clay and sulphidic peat
Profile description G (Cr1) 0–15cm	Dark grey (5Y 4/1) with few fine orange mottles; organic-rich clay or silty clay; practically unripe; mass of coarse *Phragmites* roots and rhizomes; EC_e 2 mS cm^{-1}; pH (water) 6.5, (oxidised) 4.2
Gro (Cr2) 15–30cm	Grey (5Y 5/1) clay or silty clay; half ripe; structureless; many coarse and medium dendritic pores; abundant *Phragmites* remains; EC_e 2 mS cm^{-1}; pH (water) 6.5, (oxidised) 4.2
H (Cr3) 30–80+cm	Brown peat with partly-decomposed *Phragmites* remains *in situ*; practically unripe; stinks of H_2S; pH (water) 7.0, (oxidised) 3.0 at 50cm, 1.0 at 90cm

TABLE 6.4 Soil profiles of natural wetlands: raw peat

Classification	Raised bog peat on fen peat Turbary Moor Series
Location	Holme Fen, Cambridgeshire TL 201 896
Vegetation	Birch woodland for about 90 years, formerly arable
Parent material	Peat on Oxford Clay
Drainage	Surrounding area is pump drained, higher watertable in reserve
Brief description	1.65m of raised bog peat on half ripe fen peat on Oxford Clay
Profile description 0–10cm	Litter
H1 10–30cm	Very dark brown (10YR 2/2) amorphous peat; crumb structure; abundant roots; pH 3.8

H2 30–50cm	Dark reddish brown (2.5YR 2.5/4) laminated fibrous peat with remains of *Sphagnum*, *Calluna* and *Erica tetralix*; pH 4.4
H3 50–165cm	Very dusky red (2.5YR 2/2) semi-fibrous peat with remains of *Eriophorum* and *Calluna*. Clay bands at 100 and 110cm. Below this some yellow layers of *Cladium mariscus* peat. Half ripe; pH 4.2 to 5.4
2H1 165–250cm	Very dark greyish brown (10YR 3/2.5) semi-fibrous peat, half ripe; remains of *Phragmites* and *Carex*; pH 5.5
2H2 250–280cm	Dark reddish brown (5YR 2/3) blackening on exposure to air; semi-fibrous peat to peaty clay; birch and pine remains; stinks of H_2S; pH 5.4
3Cg 280+cm	Grey (N5) clay; ripe; very firm

— Reclamation and drainage —

The natural mosaic of reedswamp, forest, carr and saltmarsh has been converted over centuries into a man-made landscape with controlled water levels, often intensively farmed and much of it no longer wetland (Chapter 7). However, some serious problems result. Drainage of peat brings an initial rapid shrinkage through loss of water (ripening): in the best documented example at Holme Fen, Cambridgeshire, the surface fell by 2m within 10 years of drainage and by a further metre over the next 30 years (Hutchinson, 1980). Drained peat in the East Anglian Fens is currently lost at a rate of about 1 cm yr^{-1} as a result of wind erosion and oxidation (Richardson and Smith, 1977), often exhuming long-buried landscapes that are now well below sea level.

Drainage also accelerates the ripening of clayey alluvium which shrinks and develops coarse vertical fissures. After many years under pasture, a dark humose topsoil is developed with a strong granular structure. The ripe subsoil, which is now only seasonally waterlogged, develops prominent rusty mottles, while the deep subsoil below the managed watertable remains grey and half ripe (Table 6.5). The differential shrinkage of the pre-shrunk and often coarser-textured levees and the buttery clay and peat in basins and backswamps exaggerates the microtopography of drained wetlands; the original basins sinking even lower relative to the levees. The more extensive tracts of alluvium, especially those of loamy texture, and fen peat lend themselves to regional drainage schemes. Most of these have been converted to arable and are no longer wetlands, although mineral soils retain 'fossil' mottles in the subsoil.

Deep drainage of wetland soils that were formerly flushed by brackish water carries a sting in the tail. These soils are often rich in pyrite, which on drainage oxidises to sulphuric acid. Where the marine alluvium is also rich in shell (such as in the reclaimed soils of the Romney Marsh area) this neutralises the acid *in situ*, and no harm is done. But sulphidic peats and clays contain little or no calcium carbonate and these develop extreme acidity. Their pH value may fall to 3 or less and huge amounts of iron and aluminium are brought into solution and carried into waterways. The aluminium is severely toxic, rapidly killing fish and invertebrates and depleting aquatic plants to a very few tolerant species. The iron precipitates as ochre in drains where the acidity is neutralised. These acid sulphate soils, may be identified in the field by straw yellow mottles of the mineral jarosite, which only forms in severely acid oxidizing conditions (compare the non-acidic profile in Table 6.5 with that in Table 6.6). Severely acid peat is not so easy to detect without measuring the reaction because it does not exhibit jarosite, but its lower layers are sometimes an inky black where the sulphates produced by oxidation of pyrite are reduced once again to iron monosulphide.

Acid sulphate soils may be managed either by heavy liming to neutralise the acidity (lime requirements in excess of 100 tonnes ha^{-1} are common) or by raising the watertable again, in order to flood the acid-generating layer and reduce the acidity once more to sulphides (Dent, 1986, p. 199).

Table 6.5 Soil profiles of managed wetlands: grazing marsh

Classification	Ripe clay with unripe subsoil (Dent, 1986)
	Mollic fluvisol (FAO/ISRIC, 1998)
	Newchurch Series
Location	Wheatacre Marshes near Aldeby, Norfolk
	TM 472 961
Landform	Flood plain of the River Waveney, subdued microtopography of creek channels, levees and basins, amplitude .5 to .75m
Land use	Meadow with watertable maintained at c.50cm in summer by pumping from open ditches. Surface water

	very common during winter. Grass species include *Phleum pratense, Cynosurus cristatus, Anthoxanthum odoratum* and *Lolium perenne*
Parent material	Calcareous estuarine alluvium
Profile description	
Ah 0–18/24 cm	Very dark grey (10YR 3/1) humose silt loam; strong coarse granular structure; friable; abundant fine and very fine roots; earthworms; pH 5.5; abrupt, irregular boundary with pockets of Ah material in fissures between the peds of the B horizon
Bg 18/24-40 cm	Grey (5Y 5/1) with common medium distinct yellowish red (5YR 4/8) mottles, silty clay; ripe; strong coarse columnar structure; firm; common white patches of gypsum on ped faces; occasional fine iron oxide nodules; abundant fine and very fine tubular pores; common fine roots; pH 6.0; clear wavy boundary
Go (Cg) 40–100+cm	Grey (5Y 3/1) with common strong brown (7YR 5/6) mottles; silty clay loam; very weak coarse prismatic structure in upper 30 cm, massive below; sticky, slightly plastic, half ripe; few fine pores, few fine roots; common shell fragments; pH at 50 cm 6.2, at 90 cm 5.9–6.2; EC of groundwater 3.2 mS cm^{-1}

Table 6.6 Soil profiles of managed wetlands: arable

Classification	Ripe clay with unripe acid sulphate subsoil (Dent, 1986) Thionic fluvisol (FAO/ISRIC, 1998) Newchurch Series, acid sulphate variant
Location	Wheatacre Marshes near Aldeby, Norfolk, 200 m south of Black Mill Pump TM 477 938
Landform	Flood plain of River Waveney. Subdued microtopography of creek channels, levees and basins, amplitude c.1 m
Land use	Arable since 1963. For the last 35 years, watertable has been held below 1 m throughout the year
Parent material	Estuarine alluvium with *Scrobicularia plana* shells below 1.3 m. Above this fresh and brackish-water deposits with transported broken shell, and patchy sulphidic layers

Brief description	A deep, stone-free silty clay. Granular humose topsoil incorporated by ploughing with mottled, ripe subsoil; on dark grey, half ripe silty clay, severely acid horizon below 70cm
Profile Description	
Ah 0–15cm	Very dark grey (10YR 3/3) humose loam; strong coarse to very coarse granular structure with clods of denser dark grey subsoil incorporated by ploughing; consistency slightly hard when dry, friable when moist, abundant fine and medium roots; many earthworm burrows and casts; abrupt smooth boundary; pH 6.5
Bg1 15–30cm	Very dark greyish brown (10YR 3/2), coarse diffuse mottles of dark brown (7.5YR 4/2) with fine cores of reddish brown (5YR 4/4); silty clay loam; strong coarse subangular blocky and blocky (1.5–3.5cm) breaking to strong medium blocky, firm; occasional fine dendritic pores within peds but most roots confined to ped faces and tongues of Ah material; earthworms; occasional white gypsum efflorescences on ped faces; pH 6.5; smooth merging boundary
Bg2 30–50cm	Dark grey (5YR 4/1) mottled dark reddish brown (5YR 3/4). Ped faces are uniformly grey; within ped mottles occur around root channels; silty clay; strong, very coarse prismatic structure 10 x 10cm section at top of horizon to 25 x 25cm at base; firm; occasional fine roots within peds but most roots confined to ped faces which bear the impressions of the root network, gypsum efflorescences on ped faces; smoothly merging boundary
Go (Cg1) 50–70cm	Dark grey (5YR 4/1) silty clay; strongly developed very coarse prismatic structure; nearly ripe; common fine tubular and dendritic pores encased in soft iron oxide pipes; broken thick soft iron oxide coatings on ped faces; occasional fine roots; abrupt smooth boundary
Gj (Cg2) 70–105cm	Dark grey (5YR 4/1) with common fine prominent sharp pale yellow (5YR 7/4) mottles and thin soft coatings of jarosite around pores; silty clay; half ripe; strongly developed very coarse prismatic structure with fissures 4cm wide at top of horizon, closing at base; many fine pores and common coarse pores with partly decomposed *Phragmites* remains *in situ*; rare live roots; pH 3.4 oxidised pH 2.0; clear smooth boundary

Gr (Cg3) 105–110cm+	Dark grey to dark bluish grey (5YR 4/1 to 5B 4/1) with many coarse diffuse black mottles; silty clay; structureless; half ripe; occasional brown soft iron pipes around partly decomposed *Phragmites* roots; many fine to coarse pores commonly with root remains *in situ*; occasional *Scrobicularia plana* shells; pH 5.5 oxidised pH 4.5.

— WETLAND SOILS IN UPLANDS —

Wetland is extensive in the uplands. Even in the dry east of Britain, stagnogley soils occur on very slowly permeable parent materials on gentle slopes. In summer, the watertable falls, but in winter it rises as evaporation is reduced. The zone of fluctuating watertable is marked by grey and rusty mottles.

Wetland is more extensive in the cooler, wetter uplands where drainage through the soil profile may be further impeded by a thin iron pan that seals the subsoil, usually at the top of a compact glacial till, and peat accumulates at the soil surface. On gentle slopes in the wettest regions, the whole landscape is mantled in acid blanket peat. Overcoming such drainage problems and acidity in upland soils was important in the development of Exmoor Forest during the first half of the nineteenth century (Chapter 12).

This chapter has outlined the soil factors, and attendant problems, associated with primary wetlands and their reclamation for agriculture. The following chapter considers the hydrological management of reclaimed lowland areas comprising both peat and mineral alluvial soils.

CHAPTER 7

Hydrological management in reclaimed wetlands

HADRIAN COOK

— INTRODUCTION —

This chapter is concerned with the hydrology of reclaimed wetland landscapes, and with their maintenance as productive farmland by the removal of excess water from them. Settlement in former primary wetland areas dates from the late prehistoric era, and subsequent human intervention has created a range of modified wetland landscapes which share many common characteristics, the most notable of which is the need for hydrological management at all scales. Reclamation commenced with the embankment of primary wetland, and with the digging and maintenance of drainage ditches. Later it involved the installation of culverts, sluices and boards in order to control water levels, and of flap outfalls set to prevent the ingress of saline waters at high tide in coastal areas. Later still came the adoption of various means of pumping, and finally the installation of underdrains.

Field-scale drainage seldom occurs without regional and 'main river' improvements. Arterial drainage schemes are concerned with the regulation of rivers (Green, 1979a) and impact at the regional scale. For example, regional drainage improvements to the Somerset Levels between the 1930s and 1970s reduced flood risk, and lengthened the grazing season by 4 months from between May and September to between March and November (WWA, 1979).

Flooding at virtually any time presents a risk to intensive crops; yet reeds and saw-sedge, both valuable crops in pre-industrial societies, grow vigorously in water 0.5m deep: hence differences in drainage requirements often presented land-use conflicts in wetland areas. Apart from surface water flood risk, freeboard drainage defines agricultural and other management options (Cook and Moorby, 1993). Grazing marsh is frequently flooded in winter, with ideal summer watertables at between 0.3 and 0.5m; for intensive grassland, levels of around 0.5m are desirable; for arable cropping, between 0.8 and 1m in spring and early summer. 'Sub-irrigation', the term used for the upflux of water to the rootzone from a shallow watertable, remains an opportunity in soils of suitable hydraulic properties (especially silty and fine sandy subsoils), where groundwater salinity presents no problems.

— RECLAIMED WETLAND LANDSCAPES —

Functionally, we may recognise three kinds of wetland. These are primary, secondary and tertiary wetlands, essentially increasing in their degree of hydrological interference which operates upon two major kinds of soil (see Chapter 2) – peat (e.g. the Adventurer's series), and the reclaimed alluvial soils (typified by the Newchurch series). However, these soil types are often intimately associated, one with the other.

Table 7.1. shows how these stages in wetland reclamation are managed, being essentially unenclosed primary wetland, at the mercy of terrestrial or marine flooding; embanked and ditched secondary; and 'empoldered' tertiary wetland. Tertiary wetland is largely under arable cultivation, maintained by deep drainage and pumping; grazing is the preferred land use on secondary wetlands, typified by the Halvergate Marshes and the Somerset Levels, as flooding is less detrimental to permanent grass. However, such distinctions are not hard and fast over the long term: Romney Marsh proper, for example, was extensively used as arable land during the early medieval period, c.1050 to 1250 (Reeves, 1995), although later was almost entirely under grass.

There is much to be learned about secondary marsh formation, development and hydrology through an examination of the pattern of ditches found in these landscapes (Cook, 1994). Figure 7.1 shows a detail of Halvergate Marshes, Norfolk. The soils are predominantly reclaimed marine alluvium (Hazleden, 1990) and their use as grazing dates from late-Saxon times (Dymond, 1990, p. 115). The landscape may have evolved between the thirteenth and seventeenth century (George, 1992, p. 236), through the cutting off of intertidal mudflats and saltings following embankment (Coles and Funnel, 1981). Saltmarsh creeks in these situations provided a ready-made system of gravitational drainage, and the sinuous nature of many present-day ditches reflects their continued use. Following embankment, a need generally arose to increase the drainage density (in this area the density is $7.7 km.km^{-2}$) and this was achieved by cutting *ad hoc* straight ditches. Other features present

TABLE 7.1 Stages in wetland reclamation

Major wetland category	Wetland form (saline)	Wetland form (freshwater)	Productivity/land use	Ecological interest
Primary	Unenclosed tidal mudflats, creeks, saltmarsh	Riverine/lacustrine, mudflat, bog, fen, carr, swamp, uncontrolled flooding	Productivity high: nutrient enrichment from flood waters and sediment; salting with grazing	High: natural habitat for flora and fauna
Secondary	Enclosed non-tidal mudflats with brackish pools and saltmarsh vegetation	Flood defences impart control over hydrological regime, limited flooding managed (i.e. flood meadows, marshlands)	Productivity reduced due to reduced nutrient and sediment inputs; leaching from rainwater becomes important; improved access improves grazing	Range of semi-natural habitats for flora and fauna, especially birds
	Unstable soft mudflats. Firm, consolidating salt mudflats (prone to rain flooding)			
	Stabilised, non-saline grassland with network of inter connecting ditches; relatively high watertable, prone to winter flooding; reedbeds associated with ditches		Productivity low without agricultural intensification; grazing marsh/hay meadows and suitable for reed cultivation under modified watertable regime	Interest high due to relict flora and fauna associated with ditches; important seasonal habitats for birds due to flooding
Tertiary	Stabilised, non-saline grassland with network of inter connecting ditches most suitable for intensive production; lower watertable, reduced flood risk; longer growing season than secondary grassland; underdrainage may be present		Productivity increased by watertable management, field levelling and fertiliser inputs; intensive grazing and silage cutting	Interest low, poor species diversification; eutrophication of water courses common
	Deep-drained, low watertable (deepened ditches) with field underdrainage, improved arterial drainage		Arable or intensive grassland	

(Source: Cook and Moorby, 1993).

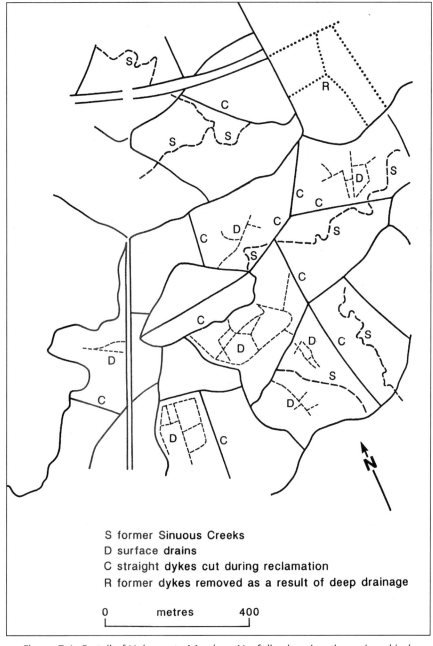

Figure 7.1 Detail of Halvergate Marshes, Norfolk, showing the various kinds of watercourse present on reclaimed alluvial marshland

in Figure 7.1 include abandoned sinuous pre-embankment drainage lines (presumably these were found too winding to be used as field boundaries); ditches abandoned or infilled following the change to underdrained 'tertiary' land uses (Table 7.1); and small linear in-field features visible from aerial photographs, most of which are small surface drains dug to reduce flooding in shallow basins. The latter result from increased soil shrinkage where finer deposition occurred away from the lines of primary creek deposition, and also from upcast from principle drains building up the perimeter of each drained area.

Figure 7.2 shows an area of early Fenland reclamation at Burwell Fen (after Taylor, 1981). The area was isolated by the cutting of the Reach Lode (Roman) with the Wicken and Monk's Lode (possibly Roman), but had otherwise remained undrained until the seventeenth century. The original field pattern was dominated by straight, parallel ditches typically in the order of 1,000ft (300m) apart and up to 1.44 miles (2.3k) long. By the mid-nineteenth century the pattern was becoming more orthogonal, with cross dykes and 'engine drains' cut in connection with the installation of pumps driven by steam or internal combustion engines between 1840 and 1940. Ditch patterns in internal peatlands tend to be more ephemeral than in alluvial soils, due to the shrinkage of peat (through de-watering, microbial breakdown and wind erosion). This would have necessitated an increased density of ditches in order to maintain freeboard, and eventually brought an end to gravity drainage, necessitating the introduction of mechanised pumping technologies.

Figure 7.3 shows the present-day pattern of drainage ditches on Tadham Moor, Somerset. Here, secondary wetland on peat soils was reclaimed in the eighteenth century and is currently managed as grassland. The pattern is strongly orthogonal, and both hydrological considerations (a wetter climate) and the need to define property boundaries (e.g. to protect turbary concessions: Williams, 1970, pp. 155, 189) on enclosure produced a drainage density around 15km.km^{-2}, twice that typical of eastern England.

— Hydrology of Reclaimed Wetlands —

Primary wetland is at the mercy of natural floods; secondary less so, although still prone to winter waterlogging and flooding. Reclamation to the secondary and tertiary stages has profound implications for watertable management; this may be demonstrated by reference to the influence of underdrainage on Walland March, near Rye in East Sussex. The soils here are dominated by the Newchurch series, a calcareous marine alluvium which has a characteristic clayey and fine loamy

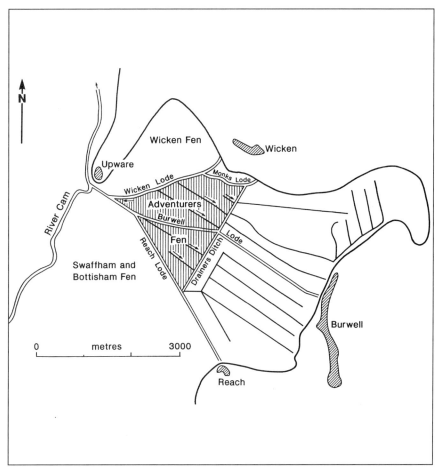

Figure 7.2 Dyke pattern on Burwell Fen, Cambridgeshire. (Source: after Taylor, 1981).

topsoil to 0.3m, and subsoils which contain thin bands of coarser material (Green, 1968).

Where the land was managed as permanent pasture (maintained under an SSSI agreement with English Nature) measurements of a dipwell, placed at a distance of 10m into the field from the ditch, were compared with the height of water in neighbouring ditches. Figure 7.4 shows the data recorded for the period from October 1990 to August 1992; the soil surface was at 2.49m OD. This shows that the winter freeboard was typically between 0.3 and 0.5m, but sometimes reached near to the surface, notably during January 1991, when standing water

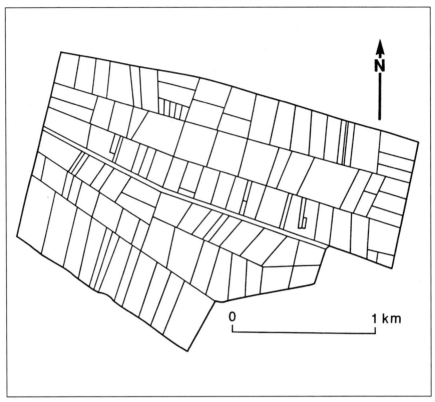

Figure 7.3 The pattern of drainage ditches on Tadham Moor, Somerset. (Source: after Cook, 1994).

collected in slight hollows in the surface of the field. Furthermore, comparison with the adjacent ditch level enables the shallow groundwater gradients to be determined. When the watertable elevation recorded in the field is above that in the ditch, subsurface drainage is towards the latter. During the summer and autumn, the situation is frequently reversed, with ditch levels above those in the dipwell. The freeboard is set at a level which permits sub-irrigation of the pasture during summer; augering at the site recorded some grass roots at below 1.0m, enabling a significant depth of soil to be exploited by the root system.

Where underdrainage has been installed on such land, on the same soil series, a freeboard of between 0.8 and 1m is maintained throughout the winter, and surface water is seldom observed in the fields. By over-deepening the ditches and main sewers, drainage from field to ditch occurs throughout the year, providing that pumping is sustained

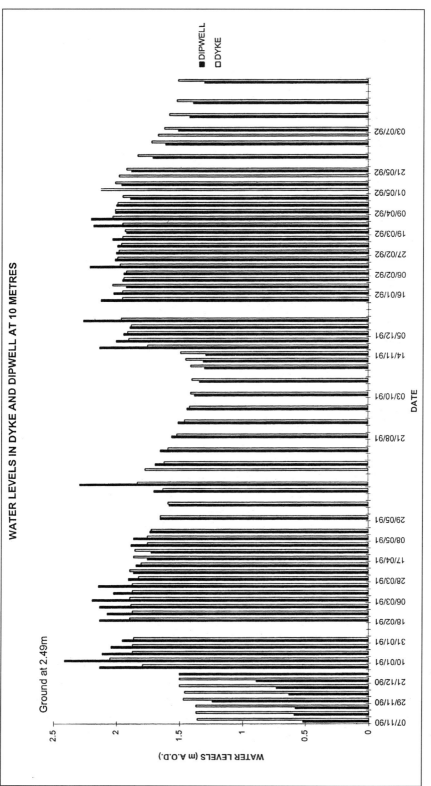

Figure 7.4 Fluctuations in watertable height in permanent pasture (without underdrainage) and adjacent ditch, Walland Marsh, East Sussex

(White et al., 1997). Compared with deep-drained sites, the watertable variations and surface flooding in secondary wetlands are thus greater, and more sensitive to week-by-week variations in site water balances.

During the late summer of 1991 a maximum soil water deficit of 85mm in the top 0.9m of soil was recorded in areas of the marsh still under grass. Values for comparable areas under arable crops, and where sub-irrigation is absent, are higher, and can range between 120 and 150mm during an average summer (Cook and Dent, 1990). Sub-irrigation of both pasture and arable may thus need to be employed to reduce the soil water deficit during the summer. In the East Anglian Fens the practice is sufficiently commonplace to represent an unlicensed and hence potentially unquantifiable draught on water resources (NRA, 1994).

Once drained to field capacity, the productivity of reclaimed marshland and fen soils is also sensitive to their available water capacities (see Chapter 2); that is, water held between field capacity and permanent wilting point. Figure 7.5 shows four soil water retention curves that relate the soil water matric potential (which may be thought of as the capillary and adsorptive 'suction-like' forces holding water within the soil matrix) to the volumetric water content of the particular soil horizon. By convention, the water content at −5kPa is taken to be that at field capacity, that at −1500kPa that at permanent wilting point. The loamy peat soil with available water capacity of 0.25 (or 25 per cent) has a much higher value than either the silty clay or silt loam. Pure peat soils typically exhibit even larger field capacities (Dent and Scammel, 1981). This in part explains the high agronomic value of peat fens, and hence the extent to which they have been reclaimed as arable, as opposed to pasture. The exaggerated curve for the 'buttery clay' (i.e., fluid subsoil) horizon from a Fen subsoil is not fully reversible were it to be re-wetted. The laboratory de-watering to which it was subjected is similar to the irreversible 'ripening' of reclaimed soils in the field.

— REGIONAL DRAINAGE CONSIDERATIONS —

Fenland or marshland ditches, which drain at a field-scale, can cope with erratic hydrological loading from adjacent fields: they tend to have high water residence times and hence display the capability to operate as a reservoir (to prevent flooding) and assist in irrigation during dry periods. Ridding an area of water also requires successful discharge to arterial water courses, and from these water must be conveyed to the sea or main river: as already noted, the ingress of

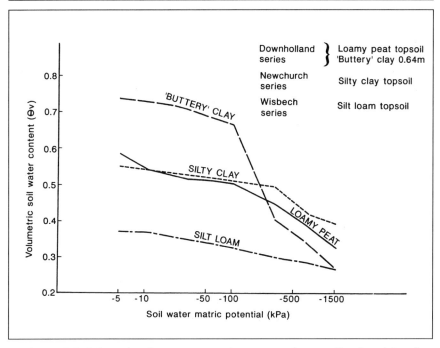

Figure 7.5 Selected soil water retention curves from soils of reclaimed wetlands

saline water could be prevented using flap sluices set in culverts at outfalls which closed at high tide. Embanking and straightening were widely practised in efforts to speed the passage of water away from reclaimed areas. In the East Anglian Fens, main rivers had low or even zero gradients, and to enhance drainage Cornelius Vermuyden created the Old Bedford river (1631) and New Bedford a Hundred Foot river (1651) as high-level carriers across the Fens to convey water towards the Wash, thereby cutting off a large loop of the Great Ouse (Godwin, 1978, p. 136). Catchwater drains are also sometimes an integral part of regional drainage schemes – as in the East Anglian Fens (Rackham, 1986, p. 383), where the Car Dyke skirts the western edge for some 140km between Cambridge and Lincoln. Built during the Roman occupation, it may have functioned as a catchwater channel diverting the flow of minor rivers away from the Fens. A modern catchwater channel has been constructed in eastern Fenland following the disastrous flooding of 1953 (Godwin, 1978). This 44km Cut-Off Channel now passes along the eastern Fen margin and has the capability to intercept flow from the rivers Lark, Little Ouse and Wissey. It re-enters the tidal Ouse below Denver. Other examples include the Royal Military canal on the northern edge of Romney Marsh.

— The development of pumping — technology

Relative increases in sea level compared with the land have been a major factor in the history of settlement and reclamation in wetlands, especially coastal wetlands; in part, these have complex geological causes, as discussed in the previous chapter. Essentially, a hiatus in sea level rise in the Roman period, and again from later Saxon times, locally combined with silting of estuaries, led to an expansion of settlement in many coastal wetlands. Land could be reclaimed and drained by gravity with relative ease, especially in eastern England (Rackham, 1986, p. 386). However, storm surges had some disastrous effects in the latter part of the thirteenth century and were, in part, a manifestation of a renewed rise in sea level. There were, at this time, some cataclysmic changes including the inundation of Old Winchelsea (the site of which was offshore from Rye, East Sussex) and the flooding of the peat workings destined to become the Norfolk Broads. In post-medieval times further changes in relative land/water levels were brought about by human agency: reclamation and cultivation of the peat Fens in East Anglia led to a progressive lowering of the land surface.

The end result of these changes was that in many circumstances gravity drainage was no longer enough to maintain areas of reclaimed fen and marsh, and in the course of the post-medieval period various forms of pump came into use, which lifted water from field dykes over an embankment and into a higher-level river or other arterial water course. In this context, the term 'pumping' is used to refer to the raising of water to facilitate land drainage; a 'pump' is, strictly speaking, a mechanical device for lifting by use of a piston, or through rotary action, but the term is used here in a rather broader sense, to include 'scoop-wheels' which lift water in a manner which may be likened to a water-wheel operating in reverse. Pumping devices require power sources. These may be human or animal, windmill-driven, steam, diesel or (most recently) electric. The employment of such technology is today essential to the success of agricultural operations in most drained areas.

The availability of a technology does not immediately mean it is adopted: it may be rejected because of unproven reliability, lack of capital for investment, and so on. Windmill technology had been available in England since the twelfth century, but – so far as the evidence goes, – it was only from the seventeenth century that it was adopted on a wide scale, initially in eastern England, to drive scoop-wheels which lifted the water from the low-lying drainage ditches into higher-level arterial water courses. Horse-drawn scoop-wheels were also sporadically employed – Bacon (1844) refers to an example at Surlingham on the Norfolk Broads – but mills were cheaper to operate, once the initial investment had been made.

An account of the development of drainage windmills is given by Williamson (1997, Ch. 5) for the Norfolk Broadlands. Here, some 72 mills survive, albeit seldom operational, and from their structure a good idea of their development can be gained. Seventeenth-century mills were always constructed of wood, but in the middle of the eighteenth century low brick towers became common: both types drove 'common' sails (i.e. sails made of canvass stretched over a wooden frame) and had a long 'tailpole' which enabled the cap to be manually hauled into the wind by the marshman (Figures 7.6 and 7.7). The sails turned a 'windshaft' which, via the 'wallower', turned a vertical axle which drove a 'spur-wheel'. This turned the 'pit-wheel', which transferred power to a horizontal axle: this turned the 'scoop-wheel'. In the middle decades of the nineteenth century, mills with more sophisticated 'patent' sails (i.e. composed of vanes that could be automatically adjusted, without stopping the mill) were erected; these were also equipped with 'fantails' which turned the cap automatically into the wind (Figure 7.8). In the late nineteenth century, 'scoop-wheels' were often replaced by turbine pumps (Figure 7.9): the same period saw a proliferation of small, light, wind pumps on trestle frames – the so-called 'hollow post mills' and 'skeleton mills'. Early scoop-wheels were around 3m in diameter, later examples could be as much as 7m or more (Williamson, 1997, p. 144): they could lift water to about three-eighths of their diameter. It was generally agreed that turbine pumps could lift half as much water again as scoop-wheels (George, 1992, p. 247).

Steam engines were employed in Broadland from the 1830s but never superseded windmills, which continued to be erected into the twentieth century, and in some cases continued in operation into the 1950s. Both began to be supplemented by internal combustion engines in 1913, but it was only the spread of the National Grid, and the widespread adoption of electrically driven pumps, which saw the demise of the older technologies. Diesel-driven pump had an output of

Figure 7.6 Herringfleet Mill on the River Waveney, Suffolk: erected in the 1820s, this is typical of early drainage mills, with its 'common' sails and long 'tailpole' to the rear.

Hydrological management in reclaimed wetlands

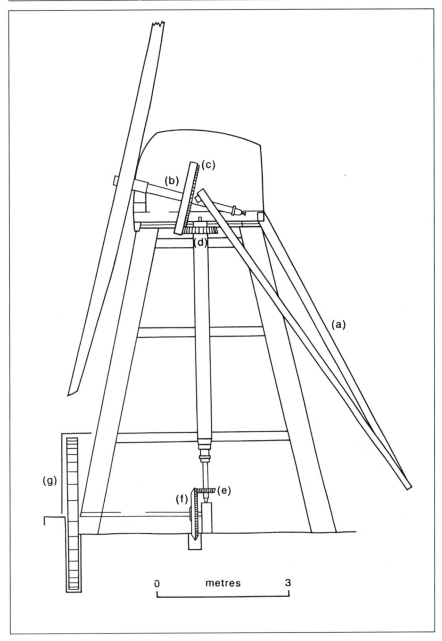

Figure 7.7 Simplified diagram of the internal workings of an eighteenth-century brick drainage mill on the Norfolk Broads. The mill is provided with common sails and has a long tailpole (a) by which it can be hauled into the wind; (b) windshaft; (c) wallower; (d) vertical axle; (e) spur wheel; (f) pit wheel; (g) scoop wheel. (Source: after Williamson, 1997.)

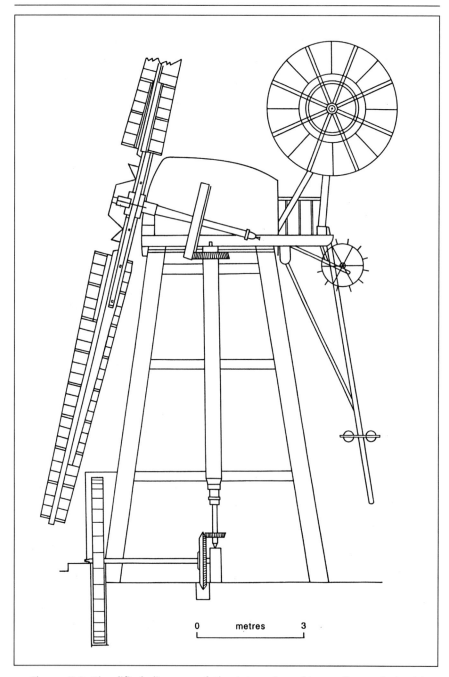

Figure 7.8 Simplified diagram of the internal workings of a typical mid nineteenth-century drainage mill with 'patent' sails and fantail (Source: after Williamson, 1997.)

Hydrological management in reclaimed wetlands

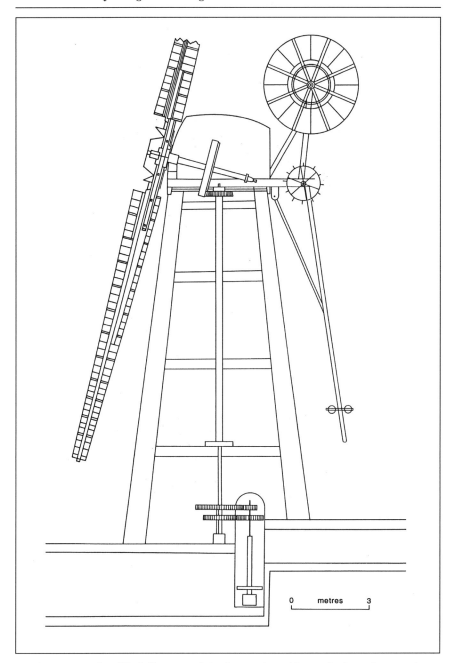

Figure 7.9 Simplified diagram of the internal workings of a late nineteenth-century drainage mill. Like the mill in Figure 7.7, it is provided with patent sails and a fantail, but it drives a turbine pump rather than a scoop wheel. (Source: after Williamson, 1997.)

35 tons of water per minute; the combined output of electric pumps raised the figure to 80 tons per minute (George, 1992, p. 251).

In the East Anglian Fens (unlike the Broads) arable rather than pasture predominated, and the soils consisted of peat rather silt or clay. Pronounced shrinkage followed drainage, and this produced a rather different history of pumping technology. Windmills driving scoop-wheels, often in series (providing a 'double-lift' up to the rivers), were more widely established and at an earlier date than in Broadland. Progressive shrinkage led to the early adoption of steam drainage, again utilising scoop-wheels. Between 1820 and 1850 the number of operational windmills fell from around 700 to about 220. Steam drainage was improved in the second half of the century by the adoption of centrifugal pumps, although scoop-wheels continued to be employed. After 1913 diesel engines driving centrifugal pumps began to replace steam pumps; but as on the Broads, only after the Second World War did electrical motors, and mixed flow or axial flow pumps automatically operated, come to dominate land drainage (Darby, 1966; Godwin, 1978, p. 140).

Not all wetland areas display these kinds of pattern: of the early adoption of wind drainage, followed by the use of steam and ultimately of pumps driven by internal combustion or electric engines. There is little evidence for the use of windmills on the Somerset Levels, although steam-driven pumps were used from 1830 (Williams 1970, p. 160). In the Dowels area of Romney Marsh windmills were only employed locally, to lift water into the embanked Royal Military Canal (Robinson, 1988); steam pumps were installed in 1852 and replaced with improved models in 1876; diesel pumps appeared in 1938, although these in turn have been replaced by electric pumps (although the magnificent diesel pumps at Appledore remain operational and are pressed into service as the pumping requirement increases).

To summarise, the development of the semi-hydromorphic landscapes typified by the East Anglian Fens, Romney (and adjacent) Marshes, the Norfolk Broads and Somerset Levels and Moors represent a triumph of environmental engineering, much of which dates from a period when scientific principles were poorly appreciated. Gravity drainage was used in the first instance, with pumping technology increasingly developed to improve the depth of freeboard drainage and reduce flood risk in the face of such problems as soil (especially peat) shrinkage, and rising sea levels.

CHAPTER 8

Romano-British reclamation of coastal wetlands

STEPHEN RIPPON

— INTRODUCTION —

Reclamation involves the construction of an embankment around an area of marshland and its subsequent drainage: this represents a major transformation of a landscape and entails the investment of considerable resources in terms of labour and resources. From the middle to late Iron Age there is evidence for small-scale drainage and settlement in a number of inland wetlands, notably on river gravels (Fulford, 1992; Lambrick, 1992; Robinson, 1992). These settlements were associated with ditched field systems that would have served a number of functions including drainage, stock management and potentially property division. In contrast, around the coast, most marshland areas appear to have been subjected to a sustained period of tidal flooding during the first millennium BC, depositing a thick layer of alluvium over the earlier freshwater peats (e.g. Rippon, 1996a, pp. 14-24; Rippon 1997a, pp. 43-4, pp. 56-60; Waller, 1994; Wilkinson and Murphy, 1995). Iron Age communities made extensive use of these landscapes, which were rich in natural resources such as fish and wildfowl, and provided opportunities for salt production and seasonal grazing. On the whole such activities had little impact on the landscape (compared to reclamation). These communities simply exploited the natural landscape rather than attempting to modify it. However, soon after the Roman conquest this started to change.

Before embarking upon an examination of Romano-British reclamation, however, it is important to consider the evidence with which we must work. There is no documentary material that is relevant to this period of wetland management and we must rely instead upon a wide range of archaeological data. Wetlands are, in fact, renowned for their excellent preservation of archaeological and palaeoenvironmental material in deeply stratified sequences of peat and alluvium (Coles and Coles, 1996), while on the surface of one of these sequences, in the East Anglian Fens, several programmes of fieldwork have revealed what is probably the most comprehensively investigated block of countryside in Roman Britain (Hall and Coles, 1994; Palmer, 1996; Phillips, 1970).

One might be forgiven, therefore, for thinking that Romano-British wetland landscapes represent a relatively straightforward subject for study, but unfortunately this is not the case. Fenland is unusual in that a large part of its Roman landsurface is no longer covered in later sediments. Elsewhere landscapes of this period are buried by up to a metre of alluvium (e.g. Essex marshes: Spurrell, 1885; Humber Estuary: Didsbury, 1988; Whitwell, 1988; North Kent marshes: Evans, 1953; Romney Marsh: Cunliffe, 1988). In addition to Fenland there are just two other coastal landscapes of this period that are not deeply buried: the North Somerset and Wentlooge Levels (both lying next to the Severn Estuary in south-west Britain). Due to the inaccessibility of Romano-British coastal landscapes elsewhere, it is to these three areas that we must turn for a detailed understanding of how coastal marshlands were exploited and managed during the Roman period.

— Options for the exploitation of — tidal marshes

There appear to have been four strategies for the utilisation of coastal wetlands during the Roman period: the exploitation of natural resources such as salt, small-scale drainage of more elevated areas, such as creek banks, followed by either piecemeal or systematic drainage.

– The exploitation of natural resources –

Wetlands support a wide range of natural resources. The coastal marshes offer the opportunity for salt production, for which abundant evidence survives around much of the southern and eastern coasts of Britain (de Brisay and Evans, 1975; Rippon, forthcoming a). The freshwater backfens were teaming with fish and wildfowl, while reeds could be used for basketry or thatch, and withies could be harvested for wattle panels. The high watertables meant that settlement would only have been possible in restricted areas of higher ground such as the fen-edge, on bedrock islands, or the banks of any tidal creeks that crossed the areas (Figure 8.1; Hall, 1992, Fig. 60; Hall, 1996, pp. 171–2, Fig. 94; Hayes and Lane, 1992, Fig. 126). With the exception of salt production and peat cutting (e.g. Hall, 1996, pp. 173-4; Hall and Coles, 1994, pp. 117-19), the exploitation of these natural resources has left few traces in the archaeological record, especially because the upper parts of the backfen peat bogs (which included the Roman horizons) have largely been lost through later peat cutting and desiccation. Rare survivals include part of a wooden platform which lay next to a small pool at Diffords, south of Meare in the Somerset Levels (Coles and Orme, 1978).

– Agriculture on a high tidal marsh –

Salt production is a seasonal activity (Bradley, 1975) and the communities involved presumably retreated to dryland settlements during the winter. However, a second strategy towards the exploitation of coastal resources involved their permanent occupation, associated with limited agriculture on areas of higher ground such as the natural banks (levees) which formed either side of tidal creeks. Fenland is a fine example with trackways, settlements and small complexes of paddocks and fields strung out along the creek banks, known locally as 'roddons' (Figure 8.1; Hayes and Lane, 1992; Palmer, 1996). Palaeoenvironmental work clearly shows that these siltlands remained as tidal marshes with salt water flowing through the creek system and artificial canals, supplying numerous salterns with their raw material (Hayes and Lane, 1992, p. 190; Leah and Crowson, 1993, pp. 44-5; Murphy, 1993a, p. 38; Murphy, 1994, p. 28; Silvester, 1988, p. 54).

It may seem odd that permanent agricultural settlements existed in such an environment but this was possible for four reasons. Firstly, the two major Fenland activities (salt production and agriculture) tended to occur in slightly different areas, with settlements on the higher creek banks towards the coast, and salterns in the lower-lying ground a little further inland (Figure 8.1). Secondly, experiments on contemporary saltmarshes in northern Germany have shown that the cultivation of cereals is possible so long as the crops are not inundated for sustained periods of time, particularly during the seedling stage (Behre and Jacomet, 1991; Van Zeist, 1974). Thirdly, the grazing of livestock is quite possible on tidal marshes: this is illustrated both by documentary material for the medieval period and by its continued practice even today, for example on the Gwent Levels in south-east Wales (Rippon, 1996a, pp. 56-7). In the case of sheep, salt water even helps to prevent certain problems such as liver fluke and foot rot which are otherwise a serious hazard in damp ground (French, 1996, p. 653; Stallibrass, 1996, p. 591). This may in fact explain the relatively

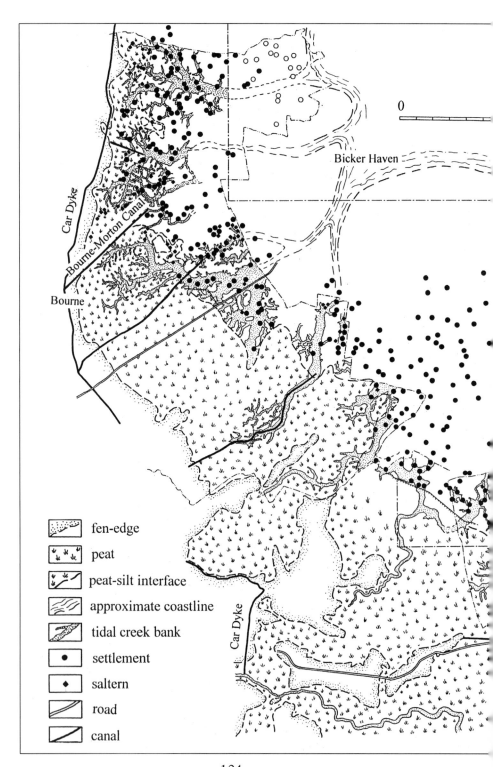

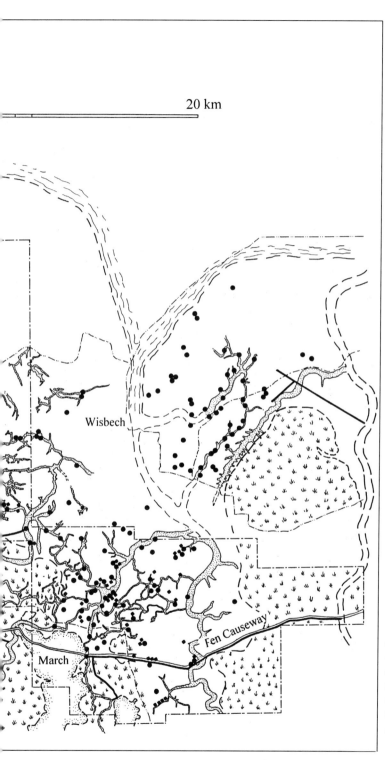

Figure 8.1 Roman Fenland. Four distinct landscape types can be recognised: an unoccupied freshwater peat fen traversed by several roads and canals; the peat-silt interface-which was used for salt production; the higher silt-land/creek banks which saw extensive settlement and localised drainage; the coastline itself (which is now buried by later sediment). (Source: based on Hallam (1970), Hall (1996), Hayes and Lane (1992), and Silvester 1988; 1991).

high proportions of sheep, as opposed to cattle, found on Fenland sites compared to other Romano-British rural settlements. Finally, during the first to third centuries AD most areas of southern Britain appear to have experienced a pause in the long-term trend of post-glacial relative sea level rise (Brigham, 1990), as a result of which the tidal marshes were subjected to less frequent inundation.

– Piecemeal drainage of a reclaimed wetland –

Palaeoenvironmental evidence from the East Anglian Fens clearly shows that the coastal siltlands were still an intertidal landscape during the Roman period: there was no sea wall to keep the tides at bay. Another coastal wetland with well-preserved Romano-British earthworks is the North Somerset Levels which have also been the subject of a long-term research project, though this work is indicating a rather different pattern of land use (Rippon, 1994a; Rippon, 1995b; Rippon, 1996b; Rippon, forthcoming b).

A combination of fieldwalking, earthwork and geophysical survey has allowed the plan of several separate Romano-British landscapes to be reconstructed, each consisting of one or more nuclei of small platforms and paddocks surrounded by slightly larger areas of open pasture. In plan these agricultural complexes are fairly similar to those of the East Anglian Fens, in that they lie adjacent to natural creek systems, and in both cases, while the layout of individual settlements and field systems shows some degree of co-ordination, there was no system of planned land allotment extending over the whole marshland area (e.g. Figure 8.3; Rippon, 1997a, Fig. 22).

However, the environmental context for the North Somerset landscapes was very different to the Silt Fens. Small-scale excavations have allowed the retrieval of a wide range of palaeoenvironmental data including pollen, plant macrofossils, foraminifera, diatoms, molluscs and small mammal bones. At the earliest site (first century AD), on Banwell Moor, this data suggests a mixed freshwater/brackish ecology but with tidal waters occasionally flowing through the ditches. By contrast, later Roman sites (third to fourth century AD) at Kenn Moor and Banwell, had a wholly freshwater palaeoenvironmental assemblage, very similar to that found in and around drainage ditches on the Levels today: this change in environment around the third century AD must

Figure 8.2 (opposite) The Somerset Levels in the Roman Period. (Source: based on Rippon (1997a; Fig. 18; 1997b) and on-going fieldwork in the North Somerset Levels (Rippon forthcoming b)).

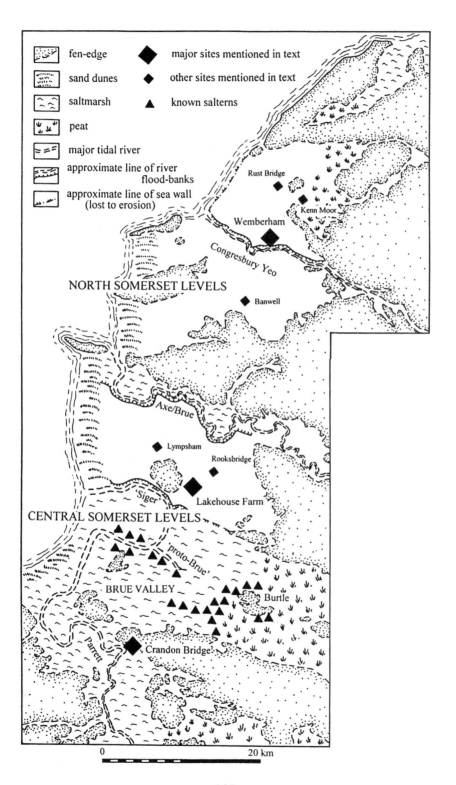

have been due to the construction of a sea wall to keep tidal waters off the North Somerset Levels.

– Systematic drainage of a reclaimed wetland –

The final strategy towards the exploitation and management of coastal wetlands in Roman Britain was their systematic and complete drainage, which only appears to have occurred on the Wentlooge Level in south-east Wales (Figure 8.4). Today, much of central Wentlooge (Peterstone parish) is divided into a distinctive system of very long narrow fields which John Allen and Michael Fulford (1986) suggested were Roman, a hypothesis since confirmed through excavation (Fulford et al., 1994). The landscape comprises a series of rectangular and trapezoidal blocks based on at least three major axial shore-parallel boundaries which can be traced for up to 5km. The land within each block is divided into a series of narrow fields up to 0.6km long, and which appear to have widths of c.16 m and c.20-21 m (Rippon, 1996a, pp. 30-2). The extent and character of this Roman drainage system is quite unlike anything else seen on other British coastal wetlands.

Only the central part of Wentlooge has this remarkable system of fields, with areas to the south-west (in Rumney parish) and north east (St Brides parish) having a far more irregular pattern, in part caused

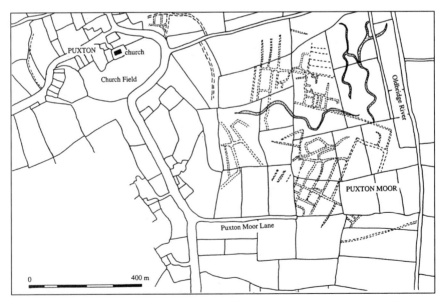

Figure 8.3 The Roman settlement and landscape at Puxton, on the North Somerset levels. Note how the Roman landscape ignores an earlier creek.

by the incorporation of meandering natural creeks that drained the former saltmarsh (Figure 8.4). This represents a very different approach to reclamation, with a gradual and piecemeal enclosure of land, contrasting with central Wentlooge and its systematic enclosure of large areas in a single episode (Rippon 1996a, Figure 8.4; Rippon 1997a, Fig. 7). So does Wentlooge illustrate a fifth Romano-British approach to saltmarsh reclamation?

Allen and Fulford (1986) suggested that both the irregular and planned landscapes on Wentlooge are of Roman date (Allen 1996). In contrast, Rippon (1996a, p. 30, 35, Figs 10 and 34) has argued that most if not all of the Wentlooge Level may have been systematically reclaimed during the Roman period but that areas adjacent to the coast and major tidal rivers were subsequently flooded, burying the Roman landscape under alluvium. When these areas were re-reclaimed a more piecemeal approach was adopted, resulting in the irregular landscape seen today in Rumney and St Brides. No known Romano-British saltmarsh reclamation bears any resemblance to the irregular patterns of fields seen in Wentlooge: in Somerset and Fenland small blocks of rectilinear fields tend to be laid out beside (or even ignoring) natural creeks (e.g. Figure 8.3), in contrast to the Wentlooge 'irregular' landscapes where the creeks themselves are assimilated into the field boundary pattern. In addition, 'irregular' landscapes such as those seen on Wentlooge are also found elsewhere on the Gwent Levels, as well as in Somerset, Romney Marsh and Fenland, and are all securely dated to the medieval period (e.g. Rippon 1996a, pp. 61–96; Rippon 1997a, pp. 142–81; Rippon, forthcoming a; Silvester, 1988).

— THE CONTEXT OF RECLAMATION —

It seems, therefore, that four different strategies were adopted for the exploitation of coastal wetlands during the Roman period: salt production (e.g. the Thames Estuary marshes), limited agriculture on a high saltmarsh (e.g. the East Anglian Fens), sea wall construction and the drainage of isolated areas (e.g. the Somerset Levels), and complete reclamation (e.g. Wentlooge). The social and economic context for these different strategies is explored in detail elsewhere (Rippon, forthcoming a), but an interim statement can be presented here. It is surely no coincidence that the largest-scale drainage operation (the Wentlooge Level) lay within the immediate hinterland of the legionary fortress at Caerleon. Large amounts of pasture and meadow (for hay) would have

Figure 8.4 The Wentlooge Level: late nineteenth century field-boundary patterns. Note the contrast between the regularly arranged fields in Peterstone with the more irregular landscapes in Rumney and St Brides. (Source: based on Ordnance Survey Six Inch Series, 1st edn.)

been essential to maintain the livestock and cavalry horses associated with such a military establishment (Fulford et al., 1994).

By contrast, the context for sea wall construction on the English side of the Severn Estuary may be found in the area's strong association with villas. Both the major reclamations in north Somerset and central Somerset have villas at their centres (Wemberham and Lakehouse Farm respectively: Figure 8.2; Scarth, 1885; Rippon, 1997b), while there is also a string of villas along the fen-edge (Rippon, 1997a: pp. 87-90). Further up the Estuary in Gloucestershire, Allen and Fulford (1990) have noted that a much greater proportion of wetlands on the eastern side of the Estuary appear to have been reclaimed compared to the western side. Here they suggest reclamation may have occurred within the context of major estates related to the Cotswold villas. Indeed, in later Roman Britain the greatest economic vibrancy appears to have been in the West, reflected in the prosperity of towns such as Cirencester and the rebuilding of villas: investment in reclamation may have formed part of this wider economic phenomena.

However, while the vast majority of the Severn Estuary Levels appear to have been reclaimed one area was not: the Brue Valley in Central Somerset (Figure 8.2). While the northern half of the Central Somerset Levels was embanked and reclaimed, allowing the construction of a villa at Lakehouse Farm, the southern half was left as a tidal marsh and used for salt production. A clear decision had been made to divide this landscape and reserve one area for its rich natural resources, notably salt, while increasing the agricultural potential of the other through reclamation (Rippon 1997a, pp. 65-77). We may once again suspect that the legionary authorities had a hand in this decision. Fulford (1996) has shown that the Second Legion based at Caerleon was involved in the exploitation of natural resources on the English side of the Estuary, while a recent study of the production and distribution of pottery in south-east Dorset has shown that this was also influenced to a considerable extent by military and/or procuratorial orders (Allen and Fulford, 1996). A major supply route by which this pottery was transported to the military establishment in south-east Wales appears to have been via a port at Crandon Bridge, on the river Parrett near Bridgwater – next to the Brue Valley. Just as a large standing army would require a considerable amount of pasture and hay for its livestock, so too would it have required large quantities of salt as a food preservative and it may have been that having decided to reclaim the coastal wetlands closest to Caerleon to supply the former, the decision was taken to reserve that part of the Somerset Levels closest to their supply base at Crandon Bridge for salt production (Rippon 1997a, pp. 121-2).

Whether or not the Brue Valley was an imperial estate will never be known, though in the Roman world salt was generally regarded as a state monopoly (Broughton, 1938, pp. 566, 799; Frank, 1927, p. 191). The authorities certainly appear to have seized certain mineral resources straight after the Roman conquest, including the Mendip lead mines and the iron resources of the Weald in Kent (Jones and Mattingly, 1990, pp. 184-92). However, there is almost no evidence for a salt industry in the Brue Valley on the eve of the Roman conquest and there is no reason why the subsequent development of the industry should not have been in private hands, perhaps under an imperial licence.

This question of imperial estates leads to the issue of the East Anglian Fenland. The traditional view is that the area formed an imperial estate: an area of unsettled land that was seized by the state, reclaimed, and used to supply garrisons on the northern frontier with grain. The major arguments that have been put forward in support of this hypothesis are as follows (based on Frere, 1987, p. 268; Jackson and Potter, 1996; Richmond, 1963, pp. 128-30; Salway, 1970):

1. As virgin land, Fenland would automatically be regarded as imperial property.
2. Salt production is likely to have been a state monopoly.
3. An inscription from Sawtry (close to the Fens) refers to 'public property'.
4. Large-scale land reclamation with drainage, artificial waterways and roads is indicative of state involvement.
5. The character of the settlement pattern, comprising farmsteads as opposed to villas, suggests that this was state-owned land leased to tenants.
6. The main settlement expansion appears to have been in the Hadrianic period, and so may relate to that particular emperor's interest in land reclamation.

Each of these arguments can now be considered in turn. The Fenland Survey has demonstrated that there was settlement in Fenland during the Iron Age, though much of it may have been seasonal and associated with salt production and pastoralism (e.g. Hall and Coles, 1994; Hayes and Lane, 1992). However, though Fenland was not virgin land, it was certainly very under-used compared both to adjacent dryland areas, and to the extent to which it was settled later in the Roman period.

Todd (1989, p. 14) suggests that areas that were brought under cultivation for the first time became *ager publicus* (land owned by the Roman state), though Salway (1970, p. 10) claims that virgin territory would have become the personal property of the Emperor (the *res privata*: Todd, 1989, p. 16). Imperial estates could also have been acquired through inheritance (for example Nero acquired the royal Icenian lands, which may have included parts of the Fens, through king Prasutagus' will), and confiscations (for example after the Boudiccan revolt and the uprising of Clodius Albinus: Frere, 1987, p. 267; Salway, 1970, p. 16).

To encourage the settlement of new lands, the late first-century Flavian emperors gave greater security to tenants. This policy was pursued in the early second century by Hadrian who in addition to reorganising Britain's northern frontier has been credited with a new wave of public building and the foundation of several new towns (Salway, 1981, pp. 185-7; Wacher, 1995, pp. 378-407). Inscriptions show that in some parts of the empire, such as Africa, Hadrian was also responsible for settling wastelands (Frere, 1987, p. 276), and it has been claimed that he may have done the same in Fenland (e.g. Salway, 1981, pp. 185-9).

The one inscription that might have a bearing on the question of whether Fenland was an imperial estate comes from Sawtry in Cambridgeshire, at the point where Ermine Street comes closest to the fen edge. It reads *public[um]* ('public property'), and inscriptions of this type usually represent the demarcation between areas of public and private land (Collingwood and Wright, 1965, No. 230).

Ever since the hypothesis that Fenland was a vast imperial estate used to supply grain to the northern garrisons it has also been assumed that the Roman state must have undertaken 'extensive and systematic drainage' (Richmond, 1963, p. 128). However, the suggestion that large-scale land reclamation implies state involvement can be dismissed: there simply was no large-scale land reclamation. There were no sea walls and the field systems and trackways appear to have grown organically, showing no sign of any overall planning or co-ordination. Work on a number of sites has also suggested that while crops were grown this was a relatively small part of the local agricultural regime, with the rearing of sheep for meat and wool being far more important (Jackson and Potter, 1996, p. 687; Potter, 1981).

There is just one major road that crosses Fenland (the Fen Causeway) though this need not indicate that the Fenland was an imperial estate: there were roads all over Roman Britain built for the purpose of military conquest, subjugation, economic development and for the efficient conduct of administration and tax collection. Similarly, the construction of a number of canals in Fenland should be seen in the context of communication rather than drainage.

Another plank in the 'imperial estate hypothesis' is the unusual settlement pattern. The traditional view has been that, although there was some Iron Age occupation on the southern islands, and a trickle of settlements from the late first century AD on the Fens proper, the major period of population expansion occurred during the Hadrianic period (e.g. Potter, 1981, p. 128; Salway, 1981, p. 189). Though the Fenland Survey has established that there was rather more pre-Roman activity in the south-west Lincolnshire Fens than previously thought, many of these sites were associated with salt production and for those which were not 'it is tempting to think of the sparse scatters of Middle Iron Age pottery... found on the levees of the creeks in the marsh as the isolated domestic debris of the herdsmen and shepherds who tended their stock on the grassy marshes during the summer' (Hayes and Lane, 1992, p. 209). It does appear that apart from in the south west Lincolnshire Fens, the initial colonisation of the silt fens was indeed in the early second century (Hall 1996, p. 169; Hall and Coles, 1994, p. 115; Hayes and Lane, 1992; Potter, 1989, p. 158; Silvester, 1991, pp. 109–10). For example, out of the sixty-nine dated Roman-British sites in the Norfolk siltlands, just one may include first-century material (Silvester, 1988, p. 90).

It is not just the explosion of settlement in the early second century that makes Fenland stand out from the rest of Roman Britain, but also the lack of villas. The view put forward by Richmond (1963, p. 130), that native peasant farmers were planted in Fenland and 'left to work the land in their ancestral fashion, [with] virtually no attempt made to convert them to Roman agricultural methods' has been remarkably enduring. Despite several major programmes of field-walking (Hall and Coles, 1994; Hallam, 1970; Simmons, 1980) there are virtually no substantial buildings known on the Silt Fens. Near Whaplode Drove church, in the southern Lincolnshire siltlands, a substantial settlement has come to light through chance finds, including an altar and a 'fair proportion' of brick and tile suggesting 'a building of some pretensions for the Fens' (Phillips, 1970, p. 302). Another unusually substantial settlement lies nearby at Gedney Hill (Phillips, 1970, pp. 302–3), while in the central Lincolnshire siltlands, around Gosberton, Quadring and Dunsby, a number of surface scatters have yielded some building stone and ceramic roof tile, along with relatively high status pottery assemblages. However, there is little to suggest that these sites were anything other than slightly above average status rural settlements (Hayes and Lane, 1992, pp. 54, 90).

Therefore, it remains true that there are no known villas on the Fens and considering the amount of air photography, field-walking, intensive farming, and the long history of interest shown in the area by professional and amateur archaeologists, it seems clear that this is a genuine absence. But need this suggest that Fenland was an imperial estate? The answer is 'not necessarily'. The lack of villas on Fenland itself, and around much of the fen-edge, might simply suggest that the area was exploited within a pattern of large private estates based at villas located on the margins of the chalk and limestone uplands that border Fenland in both Lincolnshire and Norfolk. Another dimension to the lack of villas relates to the physical environment, as it is clear that there were no Roman sea walls and that tidal waters flowed through the creeks and canals (see above). The higher roddons were sufficiently flood-free to make permanent settlement and even agriculture possible, but the absence of proper flood defences must have made all areas vulnerable to occasional inundation. It may have been this physical insecurity that discouraged the construction of villas.

The preceding discussion should have illustrated why the view that Fenland was an imperial estate is difficult to substantiate but impossible to dismiss. However, there remains one further strand of evidence: the remarkable site at Stonea on one of the small islands in the southern Fens (Jackson and Potter, 1996). This settlement comprises a grid of metalled streets associated with tightly packed timber buildings, all laid out around AD 130–50. One of the compounds defined by the rectilinear pattern of streets contained a stone building whose plan is quite unparalleled in Roman Britain. The building's wealth is demonstrated by the provision of a hypocaust, mosaic floor, painted wall plaster and the abundant use of window glass, while its substantial foundations suggest that it included a large tower perhaps 15–20m high. Jackson and Potter (1996, p. 671) quite rightly point out that 'Stonea *appears* an unusual site in many respects, but this observation must be qualified because of the lack of comparable excavated data from within the Fenland region'. However, there has been such extensive field-walking and aerial reconnaissance that it seems unlikely that other Stonea-like settlements remain to be discovered. The tower structure is clearly 'a lavish piece of "display architecture" . . . quite possibly modelled on buildings in or around Rome itself . . . As such, the whole venture smacks of officialdom rather than private initiative, and leads quite naturally to its identification as an administrative and market centre on Imperially-owned land' (Jackson and Potter, 1996, p. 688).

The preceding discussion of the East Anglian Fens should have illustrated the difficulties is establishing the social and economic context in which coastal wetlands were exploited in Roman Britain. The 'imperial estate' model has been quite rightly criticised (e.g. Millett, 1990, p. 122), but if the Roman state did indeed have theoretical rights to 'new land', and sought a monopoly over salt production, then Fenland may well have been regarded as imperial domain. The fact also remains that the sudden expansion of settlement in the early to mid second century, along with the remarkable site at Stonea, may indicate that Fenland had a rather different estate structure to the rest of Roman Britain. However, the expansion of settlements and the laying out of fields and trackways was certainly not a centrally planned exercise: the organic growth of the landscape suggests that any official control of how Fenland was exploited must have been limited to collecting rents.

— Romano-British drainage technology —

Once an area of coastal wetland has been reclaimed there remain three potential sources of flooding: tidal breaches of the sea wall, freshwater run-off from the adjacent dryland areas, and precipitation falling on the wetlands themselves. The first of these can only be countered through the construction of a sea wall, while the second and third are dealt with through a system of drainage channels.

There is not a single sea wall known to date to the Roman period surviving in Britain because most if not all of the coastlines that once had them have been lost to erosion caused by rising sea levels (Allen, 1990; Allen and Fulford, 1986; Fulford et al., 1994; Rippon, 1996a). However, there is no reason to suppose that these earliest sea walls were anything more than simple earthen embankments because lower sea levels during the Roman period meant that they could have been much less substantial than those required today (e.g. Rippon, 1997a, p. 110).

One method of dealing with freshwater run-off from the adjacent upland areas is to construct a catchwater drain along the fen-edge. There are two possible examples in Roman Britain: the Car Dykes in Cambridgeshire and Lincolnshire, and Percoed Reen in the Wentlooge Level near Cardiff. The latter is undated, though its relationship to the rest of the Roman drainage system suggests that it may have functioned as a catchwater drain (Rippon, 1996a, p. 32, Fig. 8). Though the date of the Car Dykes is clearly Roman, there has been much debate over their interpretation. The bottom of the Lincolnshire Car Dyke is not flat and the presence of several 'causeways' of uncut bedrock also means that it cannot have functioned as a canal, so favouring a catchwater interpretation (Simmons, 1979). However, the presence of banks on both sides (e.g. Hall, 1987, p. 28) would have prevented both dykes from acting as catchwater drains, and so a function as a canal to aid communication seems more likely. Even with very careful surveying, a channel this length will never be exactly horizontal and causeways such as those recorded could have acted as fixed locks, maintaining independent water levels in each section of the channel.

A number of excavations have given us a certain amount of information regarding the drainage systems that were dug on the wetlands after sea walls were constructed (see Table 8.1 and Figure 8.2). Most information has come from a number of sites where limited trial trenching has been carried out, while a range of ditch profiles has also been recorded in single sections cut through the Levels (for example, pipelines and modern drainage ditches).

TABLE 8.1 Selected field observations around the Severn Estuary revealing evidence for Romano-British drainage systems

Site	Area	Fieldwork	Reference
Banwell Moor	North Somerset	Trial trenching	Rippon, 1996b; forthcoming b
Edingworth	Central Somerset	Single section	Rippon, 1997b
Goldcliff, Hill Farm	Gwent	Single section	Bell, 1994
Goldcliff, Nature Reserve	Gwent	Trial trenching	Locock, 1996
Lympsham	Central Somerset	Single section	Broomhead, 1991
Kenn Moor	North Somerset	Trial trenching	Rippon, 1994a; 1995b; forthcoming b
Rooksbridge	Central Somerset	Single section	Russett, 1989
Rumney Great Wharf	Gwent	Single section	Fulford et al., 1994
Rust Bridge	North Somerset	Trial trenching	Hume, 1993

At Banwell Moor (third century AD) there is possible evidence that attempts were made to improve the drainage of a high saltmarsh immediately after its reclamation (i.e. the construction of a sea wall). A series of steep-sided and flat-bottomed gullies were dug, c.0.5m deep, c.0.4m wide at the top narrowing to c.0.3m at the base. The gullies were parallel and closely spaced, between 2m and 4m apart. Similar gully systems have been recorded at the Goldcliff Nature Reserve on the Gwent Levels.

Evidence from a number of sites suggests that following reclamation land was divided up and drained by ditches which in general were between 1.5 to 2.5m wide and between 0.6 and 1m deep, with a U-shaped profile (e.g. Edingworth, Goldcliff Hill Farm, Kenn Moor, Lympsham, Rust Bridge, Rumney Great Wharf phase I). Very few examples of larger ditches have been recorded, though at both Kenn Moor and Rumney Great phase II there were examples c.3m wide and 1m deep.

Larger artificial watercourses have not yet been recognised in the North Somerset Levels, though they certainly existed elsewhere. In the Wentlooge Level a number of canal-like channels which form part of the present drainage system conform to the Roman pattern of drains and are probably contemporary with them (Rippon, 1996a, Fig. 8). In Fenland there was a series of major canals, most of which ran from the fen-edge to join tidal creeks that flowed to the coast (e.g. the Bourne-Morton Canal: Hall and Coles, 1994). These may have aided the drainage of certain areas, but their primary function appears to have been to improve communications, as is most clearly illustrated by the Fen Causeway which runs east to west across the southern Fens through what at the time was an unsettled area (Rippon, forthcoming a).

This archaeological evidence for Romano-British drainage features can be compared with those employed in the historic period before the advent of mechanical pumping (Rippon, 1996a, pp. 50-4, 57-8). The surface of fields tended to be ridged through ploughing, creating corrugations very similar to ridge and furrow: there is no evidence at present for this occurring during the Roman period. In the historic landscape, water drained from these slight ridges into larger gullies known as 'gripes', which are very similar to the Roman features at Banwell and Gwent Nature Reserve. Water flowed from these gripes into field-boundary ditches (for which comparable Romano-British features have been found on various sites), and then into more substantial drainage channels known as 'reens' or 'rhynes'. Today, minor reens tend to be around 2.5m wide and 1m deep (comparable to the Kenn Moor and Rumney Great Wharf features), though the major watercourses that drain the Levels are between c.5 and 7.5m wide and 1–2m deep (Marshall, 1984; Scotter et al., 1977), for which nothing comparable has been recorded on the Severn Estuary Levels in a Romano-British context.

In the historic and contemporary landscape water flows through the reen system into major rivers that discharge their waters under the sea wall through tidal doors known as 'gouts'. None have been recorded from Roman Britain (largely because the Roman sea walls have been lost), though very similar culvert-like structures, constructed out of hollowed tree trunks, have been recorded in the Netherlands and dated to the first to third centuries AD (Van Rijn, 1995).

– Conclusion –

Though evidence for the exploitation and management of Romano-British coastal wetlands is widespread, many landscapes are buried by later alluvium or have been destroyed by development, agriculture, desiccation and erosion. However, three well-preserved and extensively studied landscapes – the Gwent and Somerset Levels, and the East Anglian Fens – do afford an insight into the Roman approaches towards wetland management.

There appear to have been four strategies for wetland exploitation in Roman Britain: salt production, small-scale agriculture on a high saltmarsh, sea wall construction and the establishment of individual settlements and field systems, and large-scale reclamation including the laying-out of planned field/drainage systems. In areas that were reclaimed it seems that the techniques employed were very similar to those of the medieval period, with a hierarchy of drainage channels feeding water through tidal sluices under the sea walls.

The three agricultural strategies that are the concern of this paper were carried out in the context of at least three social and economic systems (Rippon, forthcoming (a)). The large-scale reclamation of Wentlooge appears to have been undertaken by the military authorities, while the embankment of many of the other Severn Estuary Levels may have occurred in the context of villa estates. In contrast, Fenland may have been part of either villa or imperial estates, but with its colonisation carried out by peasant communities leasing small parcels of land.

Three Romano-British landscapes have been used extensively in this study (Fenland, North Somerset and Wentlooge), and Wentlooge in particular is a nationally important monument to human achievement in wetland management: possibly the first major drainage operation in Britain, and certain the only Roman system to have survived in use in this country. Any cultural landscape in a wetland context is far better preserved than dryland examples due to the permanently high watertables and resulting waterlogged conditions, which have also led to these areas escaping the ravages of urban development and intensive arable farming.

However, Fenland shows how quickly times can change. When the Romano-British field systems were first recorded through air photography in the 1930s they survived as extensive complexes of earthworks. Indeed, Christopher Hawkes described the Roman landscape as 'so complete... that there is confident hope that a complete air survey of the Fens may give us something like an Ordnance Survey of this piece of Roman Britain' (cited by Hallam, 1970, p. 25). However, by the 1950s Sylvia Hallam was able to field-walk large areas as the post-war intensification of agriculture led to the ploughing up of ancient pasture, and by the 1980s the areas of Romano-British landscape that survived as earthworks were tiny islands in a sea of corn.

Wentlooge and the North Somerset Levels have so far escaped the unnecessary excesses of modern agriculture and remain largely pastoral landscapes. However, both areas have suffered extensive losses due to urban/industrial development, a remarkable concentration of road building, and a peculiar obsession with digging amenity lakes (in a landscape which already has plenty of water!) (Bell, 1995; Rippon, 1995a). A study in 1995 established that 39 per cent of the Gwent Levels had been destroyed, with proposed developments threatening a further 8 per cent (Rippon, 1996a, Table 9), though this figure is already an under estimate. The pace of development shows no sign of slowing, with two further major road schemes on the horizon, and unless there is a dramatic change in planning policy for this particular area of field, marsh and meadow, its history of water management may well be coming to an end after some 1,800 years.

CHAPTER 9

Medieval reclamation of marsh and fen

ROBERT SILVESTER

— INTRODUCTION —

Reclaimed wetlands offered potentially rich rewards to the medieval communities that lived on and around them. Arguably there are no comparable geographical zones on which medieval people and their successors in later centuries left such a profound imprint, creating what are, literally, new landscapes. Sir William Dugdale's indispensable volume on the History of *Inbanking and Drayning of Divers Fens and Marshes* (1662) marks the beginning of scholars' fascination with the winning and losing of land in the low-lying coastal regions of England and Wales, an interest which has continued to the present day. More recently, Michael Williams's synthesis on medieval marshland reclamation drew heavily on the Fenland research of H. C. Darby and H. E. Hallam, his own work in the Somerset Levels, and more restricted studies of the Humber wetlands, Romney Marsh and the Pevensey Levels (Williams, 1982). Almost all of this work was based on documentary material, but Williams in his earlier study of the Somerset Levels was quite specific in highlighting the landscape itself as a source of information that merited close attention (Williams, 1970, p. 256). Since 1982, the pace of research into the historic development of wetlands has quickened. English Heritage has inaugurated studies in the East Anglian Fenlands, the wetlands of north-west England and most recently the Humber wetlands; the Romney Marsh Research Trust has emerged to promote a diverse range of research into the most important of the wetlands in south-east England; and the levels on either side of the Severn Estuary are the subject of ongoing research. Few of these major studies have been geared specifically to the medieval era: in design they are multi-period and are the more important because of it. While studies based on archival material still appear regularly, much modern research reveals a more acute focus on the topography of the wetlands and the physical manifestations of drainage and reclamation; this approach is best exemplified in the work by Stephen Rippon on the levels bordering the Severn Estuary.

The wetlands of England and Wales, most of which are coastal, divide into two basic categories: those where the surface deposits are of marine or riverine alluvium; and those of organic peat. This division, however, simplifies what is frequently a complex depositional sequence where peats and alluvium interleave, leading to large tracts of wetland where a thin layer of peat may overlie alluvium or vice versa. Largest of these are the Fenlands of eastern England extending over more than 2,000 square kilometres (Figure 9.1). Around the Wash marine silts and clays create a broad coastal zone through Lincolnshire, Cambridgeshire and Norfolk, behind which freshwater peat infills the remainder of the fenland basin. Next in size are the Somerset Levels on the western side of the country, some 603km^2 of low-lying ground (Williams, 1970, p. 14) with inland peat fens and coastal claylands edging the Severn Estuary. Further north, the wetlands around the Humber again have coastal marshlands with peat inland along the Hull Valley and on the Thorne and Hatfield Moors (Sheppard, 1958; Van de Noort and Ellis, 1997, p. 10).

Some coastal marshlands have surface deposits composed almost solely of alluvium. Romney Marsh, an umbrella term for the marshes of Romney, Walland and Denge, extends over 270km^2 on the Sussex/Kent border (Eddison and Green, 1988, p. 1), its marine alluvium coating underlying peat. Facing the Somerset Levels across the Severn Estuary are the Gwent Levels, smaller and entirely of alluvium spreading across a coastal plain which, at its greatest extent, is less than 6.5km wide and some 111km^2 in area. Several of the river estuaries in southern England also fall in this category: the Thames, the Essex rivers, the Sussex Ouse and the Pevensey Levels. Further north are the marshlands along the Lancashire coast and the Solway to the west of Carlisle (Miller, 1988, p. 247).

Finally there are inland wetlands such as Otmoor in Oxfordshire (Bond, 1981) and the peat mosses of Cheshire, Shropshire and Lancashire (Middleton et al., 1995; Leah et al., 1997). These, however, were largely unaffected by the medieval boom in land improvement and are not considered further here.

— METHODS OF DRAINAGE AND RECLAMATION —

The sequence by which a coastal wetland was initially colonised and protected, and subsequently reclaimed, is implicit in much of what has been written on the Fens and the Somerset Levels, and has been conceptualised by Rippon for the Severn Levels (1996a, p. 8) (Figure 9.2). The introduction of permanent settlement, which probably evolved from seasonal or industrial activity such as grazing and salt-making, was followed quickly by the construction of a sea wall to prevent

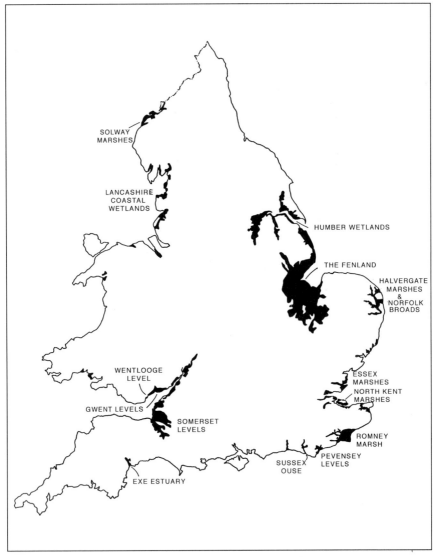

Figure 9.1 The main coastal wetlands of England and Wales referred to in the text.

marine inundation; indeed it may even have been a necessary precursor. Fledgling settlements on the higher silts were encompassed initially by irregularly laid out fields edged by drainage ditches, but as the settlements required more land, banks were constructed inland to protect new intakes from freshwater flooding. Drainage patterns within the new intakes became increasingly regular, though it should be emphasised that land might be enclosed for pasture but not drained until considerably later. On the seaward side, the sea bank would act as a buffer against which marine alluvium accumulated, raising the level of the salt marsh and creating the environment in which a further sea bank could be built that enabled more reclamation, again with regularly laid out drains. Later enclosures tended to be larger, but all were of necessity completely encompassed by embankments to form 'watertight' compartments.

In the inland peat fens, the perceived pattern was generally simpler. Most settlements, whether village or monastic establishment, occupied higher ground on the fen edge or on an island protruding from the fen. Reclamation where it occurred usually involved drainage ditches and embankments to enclose specific blocks of land, though settlement in the fen was generally limited. As on the marshlands these intakes could lead in time to an aggregated pattern of enclosures spreading out from a central locus.

The techniques developed to construct banks and dykes are not often considered, perhaps because they are infrequently recorded in medieval documents. An exception are the details of bank building on the Canterbury Cathedral Priory estates at Ebony and Appledore on Romney Marsh, both for the materials used, which included timber, straw and stones, and the techniques adopted (Smith, 1940). More usually banks were of little more than clay and earth but in the Fens tussocks of grass ('hassocks') were used to stabilise external faces, Hassock Dyke being erected in Holbeach and adjacent parishes between 1190 and 1195; and in the peat fens the banks might be planted with trees, principally willow (Hallam, 1988, p. 498). On the Pevensey Levels banks were thrown up between parallel ditches (Dulley, 1966, p. 30). Timber was used extensively, for sluices (Rippon, 1996a, p. 70), for gutters to channel streams (Smith, 1940, p. 36), and in the form of hollowed tree trunks for culverts to discharge water beneath the sea banks (Hall and Coles, 1994, p. 145). Fields were frequently divided up into narrow, regular strips separated by drainage dykes, though there were exceptions (Hallam, 1965, p. 150; Rippon, 1996a, p. 40; Rippon, 1997a, p. 149).

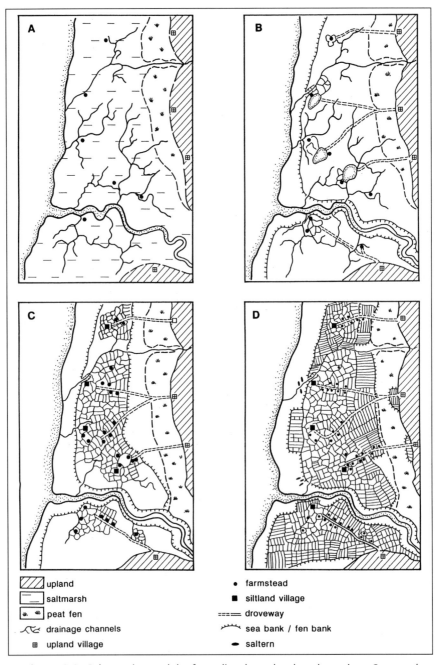

Figure 9.2 Schematic model of medieval wetland reclamation. Seasonal settlements in the saltmarshes (A) were succeeded by permanent settlements protected by a sea wall (B). Reclamation led to the erection of the first fen banks inland (C), and later most of the marsh and some of the fen was enclosed, together with younger marsh on the seaward side of the sea wall (D). (Source: based on Rippon, 1997a, Fig. 7.)

— RECLAMATION OF THE MARSHLANDS —

The utilisation of the coastal marshlands commenced early in the medieval period, despite the widespread marine flooding evidenced in the later Roman and early post-Roman centuries from Romney Marsh (Cunliffe, 1988, p. 83), the Severn Estuary (Rippon, 1997a, p. 178) and in parts of the Fens (Waller, 1994, p. 79), which had forced earlier colonisers to abandon their lands. Saxon charters reveal that the silts of the Somerset Levels were settled some centuries prior to the Norman Conquest (Rippon, 1997b, p. 175) and by the end of the Saxon era sea and river walls were in place to prevent marine flooding (Rippon, 1994b, p. 242). Substantive evidence comes from Romney Marsh proper where scatters of surface debris from the tenth century, if not earlier, indicate colonisation (Reeves, 1995, p. 82), confirming the picture derived from pre-Conquest charters and the Domesday survey (Brooks, 1988). But the most emphatic evidence currently comes from the East Anglian Fenland. The debris of middle Saxon settlement (mid seventh century to late ninth century) in the southern Lincolnshire marshlands signals extensive activity, though unrelated to the medieval and modern settlement pattern (Hayes and Lane, 1992, p. 215). A similar picture appears in the Norfolk Marshland, the coastal siltlands adjacent to the Wash, yet Late Saxon occupation is evidenced from the vicinity of existing villages, and it is here that the irregular field patterns are best observed (Silvester, 1988, p. 160). It is now generally accepted that the great sea bank, known misleadingly as 'Roman Bank', which extended around all the Wash siltlands is late Saxon in origin (Hall, 1996, p. 185). A pre-Conquest expansion inland towards the peat Fens is less easy to demonstrate, but from documentary evidence Hallam was able to deduce at least one fen bank that protected intakes in each of the Lincolnshire siltland wapentakes prior to the eleventh century (Hallam, 1965, Fig. 1); and for one northern marshland parish, Wrangle, a succession of fen banks has been claimed for the pre-Conquest period (Lane, 1993, p. 86).

It is after the Conquest that the evidence for continuing reclamation becomes abundant not only for the three main areas, but for other coastal wetlands too. Hallam estimated that in the Lincolnshire wapentake of Elloe alone, some 130km^2 of inland marsh and fen were reclaimed between 1170 and 1241 (Hallam, 1965, p. 39), and in the Norfolk Marshland about 115km^2 were drained by the fourteenth century (Figure 9.3). A succession of banks was thrown up to take in new land; the adjacent communities of Whaplode, Holbeach and Fleet in

Figure 9.3 The Norfolk Marshland landscape. The primary settlements lay close to the Wash coastline and were surrounded by irregular field patterns of late Saxon date. Post-Conquest reclamation pushed southwards, the fields becoming ever more regular in their layout. In the south was West Fen, the common grazing land of the Marshland villages, which was linked to the villages by wide droveways. (Source: Silvester, 1988; drawn by Margaret Mathews, © Norfolk Field Archaeology Division.)

Elloe built four banks between c.1160 and 1241, their alignments signalling that these were co-operative efforts by the three villages (Hallam, 1965, p. 38). In Marshland a similar picture emerges with one community, Walsoken, displaying five fen banks in the Saxon and medieval periods (Silvester, 1988, p. 86), but differing from Elloe in that each vill appears to have expanded at its own speed, erecting a new fen bank to protect only its own land, and protecting the flanks by extending the 'wardikes' that spaced it from its neighbours (Silvester, 1988, p. 164). New enclosures could be completed on a massive scale: over 23km^2 of Wisbech High Fen were enclosed at one point in the thirteenth century and divided into narrow strips that were up to 1.6km in length (Hall, 1981, p. 43).

Less significant was the reclamation of the Wash coastal marshes (Darby, 1983, p. 10). Only limited intake of the saltmarshes in Cambridgeshire and Norfolk seems to have been attempted, generally no more than one new field per community (Hall, 1981, p. 47; Silvester, 1988, p. 164), though the possibility that other reclamations outside the sea bank were subsequently inundated and lost without record cannot be discounted. In contrast, along the Lincolnshire coast considerable effort was expended in taking new land from the sea, particularly in the twelfth century (Hallam, 1965, p. 91).

In the Somerset Levels isolated fragments of evidence point to on-going reclamation, and the increase in the relative wealth between 1086 and 1327 of the coastal claylands points to medieval expansion just as in the Fenland (Rippon, 1994b, p. 246). But there is much less documentary evidence and also less detailed topographical analysis to provide a framework comparable to Hallam's research in Lincolnshire. One exception is Sowy island where the Rivers Tone and Cary joined the Parrett, creating a spread of alluvial soil. Williams (1970, p. 50) calculated from the Glastonbury Abbey records that by the mid-thirteenth century more than 390ha of low ground surrounding the island had been converted into meadow and about forty houses built there.

The Gwent Levels on the opposite side of the Severn were not settled until the late eleventh century at the earliest (Rippon, 1996a, p. 38), though one tract, the Wentlooge Level, retained a Roman drainage system which was brought back into use (Rippon, 1996a, p. 90). Defended by a sea bank which was abandoned in the fifteenth century when there was a contraction of the reclaimed area (Rippon, 1997a, p. 242), the Gwent Level communities on the higher ground by the coast expanded their lands by intakes that became larger and increasingly regular in layout as time passed (Rippon, 1996a, p. 123), though some of the low-lying backfen remained as common into the post-medieval era (Rippon, 1996a, p. 95).

Along the south coast of England from the Devon Exe eastwards (Parkinson, 1980, p. 20), reclamation of one form or another transformed low-lying estuaries and river valleys. The largest marshland area was Romney Marsh. No overall assessment of its medieval exploitation, comparable with those for Lincolnshire and the Severn Estuary Levels, has yet been published, though a preliminary overview (Cunliffe, 1980) has been succeeded by a number of papers detailing developments in specific areas of the Marsh (Tatton-Brown, 1988; Gardiner, 1988 and 1995; Vollans, 1995). Walland Marsh, lying behind the Dungeness shingle, witnessed considerable reclamation from the twelfth century, with 'innings' (intakes) continuing into the fourteenth and fifteenth centuries (Tatton-Brown, 1988, p. 110). Further west, where the Brede valley entered the Marsh, land was enclosed in the twelfth and first half of the thirteenth century by local entrepreneurs (Gardiner, 1995, p. 130). Near Lydd, marsh was intermittently enclosed at the end of the eleventh century into the twelfth century and again in the early fourteenth century (Vollans, 1995, p. 120). Further west in Sussex the estuarine marshes of the River Ouse were enclosed to provide valuable meadowland (Brandon, 1971b, p. 96), and the saltmarshes of the Pevensey Levels were 'inned' from the early twelfth century, with such celerity that most of the available marshland had been reclaimed by 1287 (Dulley, 1966, p. 32).

Documentary sources point to marshland in North Kent and on both sides of the Thames estuary being embanked and drained immediately after the Norman Conquest: around Gravesend and also in the Wantsum Channel that separated Thanet from the rest of Kent (Whitney, 1989, p. 33), though it is likely that this activity began earlier, in the late Saxon period. Canterbury Cathedral Priory reclaimed islands at the mouth of the Medway between 1279 and the beginning of the fifteenth century, and enclosed parts of the Cliffe marshes in the 1290s (Hallam, 1981, p. 76). On the Essex shore, some marshland enclosure is documented in the medieval era (Wilkinson and Murphy, 1995, p. 208) and Foulness Island was enclosed and over 15 per cent of it turned over to arable before 1420 (Rackham, 1986, p. 387). Further north in eastern Norfolk the grazing lands of the Halvergate Marshes required the protection of embankments and drainage dykes in the later thirteenth and fourteenth centuries (Williamson, 1997, p. 49).

Around the Humber some of the major rivers such as the Ouse may have been embanked by the eleventh century (Dinnin, 1997, p. 21) and during the twelfth and thirteenth centuries banks beside the Rivers Hull and Humber were joined to form a continuous barrier against marine flooding (Sheppard, 1958, p. 2). Both saltmarshes beside tidal rivers such as the Ouse, and the waterlogged low-lying clays and silts behind them were drained for pasture, meadow and some arable (Sheppard, 1966, p. 15). On the north bank of the Humber between the Hull valley and Spurn Head a strip of embanked marshland up to 5km wide was populated with farms and granges during the twelfth century, but in the later Middle Ages perhaps half of this ground was lost to the sea, in part due to changes in the landform at Spurn Head (Sheppard, 1966, p. 6). Bordering the Irish Sea, coastal marshlands in Merseyside and Lancashire as far north as Furness were embanked and used for grazing (Hallam, 1981, p. 184; Cowell and Innes, 1994, p. 210).

— Reclamation of the peat fens —

The medieval exploitation of the peat fen contrasts with the activity on the marshlands. The Lancashire and Cheshire mosses were left largely in their natural state during the Middle Ages, their resources valued sufficiently to protect them from improvement (Cowell and Innes, 1994, p. 179; Leah et al, 1997, p. 154). The same is true of the carrs higher up the Hull valley (Sheppard, 1958, p. 4) and also of vast tracts of peat fen in the East Anglian Fens and the Somerset Levels, which provided valuable grazing for communities living on the coastal marshlands. South of the Norfolk Marshland lay West Fen (Figure 9.3), about 35km^2 of peat with some alluvial silt, which was the subject of an agreement in 1207 between the major landowners of Marshland to retain their rights to pasture and turbary whilst excluding adjacent communities. West Fen was already partitioned from the Fens to the east by Chancellor Dyke, and in 1223 the Old Podike Bank was raised along its southern boundary to keep out freshwater floods from the south (Silvester, 1988, p. 32). It survived as common pasture until the late eighteenth century.

There is, however, no implication here that the peat fens were fervently avoided; rather that improvements, particularly in the East Anglian Fenland and the Somerset Levels, were initiated on a piecemeal basis and usually where the peat deposits were relatively shallow (Darby, 1940, p. 43; Williams, 1970, p. 41). Only the deeper fens were eschewed (Hallam, 1965, p. 91). At Bourne in south Lincolnshire three phases of medieval reclamation are visible in Gobbolds Park a kilometre to the east of the town, but these intakes represent only a very small part of Bourne Fen (Hayes and Lane, 1992, p. 141). The abbot of Thorney (Cambridgeshire) built a house and offices in Thorney Fen, surrounding them with ditches, and also enclosed a large part of the fen, anticipating that the ground might be turned to arable or meadow in due course (Darby, 1940, p. 50).

The peat fens of the Somerset Levels were almost certainly used for grazing in the Saxon period but it is generally agreed that drainage and reclamation of the moors did not occur on any scale until the thirteenth century (Williams, 1970, p. 17, Fig. 6; Rippon, 1997b, p. 179). Again it was the shallower peat overlying alluvial deposits or mineral soil that was singled out, as around the periphery of King's Sedgemoor (Williams, 1970, p. 74). Instructive in this context are the earthworks of a relict landscape at Huntspill, where drainage ditches extend off the alluvial clays into the peat, generally for only a short distance but exceptionally for over 1.6km (Rippon, 1997b, p. 210). But the absence of references to draining and reclamation in the abundant documents from Wells, Glastonbury and the other religious establishments with extensive landholdings in the Levels underlines the limited extent of medieval activity here (Williams, 1970, p. 46).

Marshland and peatland alike were both served and threatened by the natural streams and rivers that flowed across them. These watercourses provided arteries for transport and communication and functioned as drainage channels. At the same time they were the conduits for the freshwater floods that covered the reclaimed lands, and their estuaries might funnel marine floods into the coastal marshlands. It was inevitable that for one reason or another medieval communities, particularly the monastic ones, would attempt to modify the waterway systems of the wetlands, particularly in the peat fens. Cnut's Dyke was built as early as the tenth century to aid the construction of Ramsey Abbey (Hall, 1992, p. 42), and it has been said of the Fens generally that it would be difficult to find a river whose course was not to some extent artificial (Owen, 1981, p. 42). Both the Nene and the Great Ouse, the two major rivers of the southern Fens, were diverted and canalised at different times in different places. The former was realigned over 11km at March (Cambridgeshire) in the late Saxon period (Hall and Coles, 1994, p. 136), the latter was diverted to serve the monastery and cathedral at Ely before the twelfth century (Hall, 1996, p. 40). Many rivers, such as the Shire Drain near Thorney, were embanked to prevent flooding of adjacent land (Hall, 1987, p. 52). Fen-edge settlements in Cambridgeshire and Norfolk frequently had inland ports or landing places (hithes) linked to the Great Ouse system by canals, most of which can be still be detected today (Silvester, 1993, p. 34; Hall and Coles, 1994, p. 137).

A similar trend appears in the Somerset Levels where Glastonbury Abbey constructed a canal as early as the tenth century (Rippon, 1997b, p. 212), and in later centuries the courses of major rivers such as the Parrett and Brue were altered, and the Axe was also canalised with hithes at various points (Williams, 1970, p. 53; Rippon, 1997b, p. 213). New watercourses were imposed on the landscape, such as the Pilrow Cut which improved the flow of the Brue to the west of Wedmore Island but was also a means of linking Glastonbury with its coastal manors in the thirteenth century, facilitating the movement of produce and materials (Williams, 1970, p. 68). Elsewhere a massive channel known as the Rhee Wall was built across Romney Marsh in the thirteenth century to serve New Romney (Vollans, 1988). Meaux Abbey in the peatlands of the Hull valley cut a number of canals in the twelfth and thirteenth centuries: Eschedike provided a waterway from the River Hull direct to the abbey itself, while Skerndike served a grange to the north of the abbey (Sheppard, 1958, p. 3).

— The reclaimers —

Not without some justification is it frequently assumed that the monasteries and other religious establishments played the major role in the medieval improvement of the wetlands. Monastic records are relatively abundant, offering written testimony of drainage effort from before the Conquest to the earlier sixteenth century, and where they are absent interpretative problems emerge (Hallam, 1965, p. 107). In some regions ecclesiastical control was almost total. Monastic establishments had been founded on the edge of, or even on islands in, the more extensive wetlands during the later Saxon era. Some became famously large and wealthy, their prosperity dependent on the acquisition of land through grant or purchase, and on its successful exploitation. The abbeys of Glastonbury, Muchelney and Athelney, together with the bishopric of Wells, commanded most of the Somerset Levels (Williams, 1970, Fig. 4), and much of the improvement will have been conducted directly by them or indirectly through their tenants. Nevertheless, during the twelfth and early thirteenth centuries Glastonbury's control over its landholdings relaxed, leading to illicit improvement on the low-lying moors. Later abbots expended considerable effort regaining control of these intakes (Williams, 1970, p. 39).

The peat fens of Cambridgeshire were dominated by the bishopric of Ely and important monastic houses, such as Thorney, Ramsey and Crowland. Only one estate in the south-western Fens – Holme – was in lay ownership (Hall, 1992, p. 103), and it can be assumed that most of the intakes were monastic in origin. A similar picture emerges for the fens along the Lincolnshire Witham with Bardney, Tupholme and Kirkstead Abbeys amongst others playing a significant role in changing the appearance of the landscape (Hallam, 1965, p. 99). On the Gwent Levels, Tintern and Llantarnam Abbeys and Goldcliff Priory all had substantial holdings, as did at least two Somerset abbeys (Rippon, 1997b, p. 218). More than 40 per cent of Romney Marsh proper was owned by the archbishop and cathedral priory at Canterbury, with further large estates belonging to other monasteries in Kent (Brooks, 1988, p. 90). This picture is repeated in virtually every wetland area of England and Wales: Canterbury reclaiming land in the marshes of North Kent, Rievaulx Abbey in the Vale of Pickering (Miller, 1988, p. 249), Meaux Abbey in the Hull valley (Sheppard, 1958, p. 3), Selby Abbey in the Humberhead Levels (Dinnin, 1997, p. 21) and Merevale Abbey draining Altcar in Lancashire (Farrer and Brownbill, 1907, p. 223).

However, much reclamation that can still be recognised from ground evidence was never documented and the role of secular landholders in these operations should not be dismissed or underestimated. Some wetland zones such as Wentlooge in the Gwent Levels were primarily in the hands of secular lords (Rippon, 1996b, p.79), and it is in the coastal marshlands where the initiative of freeholders, tenants and secular lords is most in evidence, though paradoxically it is the monastic charters that provide confirmation of this. The evidence points to the co-ordination of communities, sometimes even larger groupings, in the construction of fen and sea banks, both in the Fens and the Somerset Levels (Hallam, 1965, p. 19; Rippon, 1997b, p. 180). Once enclosed it was individuals who undertook drainage and improvement. In 1286 Thomas of Holbeach reached agreement with the communities of Holbeach and Whaplode for the construction of a sea bank on his land, the leading men of the villages acting as signatories (Hallam, 1965, p. 16). Earlier, in the late twelfth century in Holbeach, Conan son of Ellis was active as an improver, most of the records relating to his works being monastic charters (Darby, 1940, p. 46; Hallam, 1965,

p. 10). Monasteries were involved on the coastal siltlands but less universally than in the peat fens. Blocks of lands in the new fenward reclamations in Elloe wapentake were granted regularly to monastic houses such as Spalding and Crowland, but the monasteries can be recognised as latecomers, acquiring already reclaimed land through grant and purchase in order to aggregate enough ground to establish viable granges, of which there were a substantial number on the Lincolnshire marshlands (Hallam, 1965, p. 130; Map 6). It is a pattern that is repeated elsewhere around the Wash. The Lewes Priory cartulary presents a similar picture for the Norfolk Marshland (Bullock, 1939), and the agreement of 1207 regarding the common of West Fen lists the main landholders in Marshland. Ely and Lewes head the list, but no less than fifty secular landholders are named, several of whom appear elsewhere in the charters as donors of land to Lewes (Bullock, 1939, p. 67). Such records offer only a fragmented and indirect view of how reclamation developed, but are certainly representative of wider trends.

Domesday Book has been used to argue that lay manors in the Somerset Levels underwent greater improvement than their ecclesiastical counterparts (Rippon, 1994b, p. 242). And the emphasis could change through time. Battle and Robertsbridge Abbeys were actively reclaiming land in Romney Marsh by the end of the twelfth century, but they seem to have been preceded by local improvers about whom little is known (Gardiner, 1988, p. 114), and in the Brede Valley it was local men who were active in the early thirteenth century (Gardiner, 1995, p. 130).

What emerges is that much reclamation went undocumented, and it is rare to encounter records of intakes by local communities and their secular lords. The monasteries undoubtedly played a major role but as Rippon (1997b, p. 218) has stressed the character of the surviving documents can exaggerate the importance of their contribution to the overall picture. While individuals undoubtedly played a part in instigating reclamation, the communal effort that went into constructing, and subsequently maintaining and repairing the banks and dykes, often through the immutable factor of manorial service, needs to be emphasised (Salzmann, 1910, p. 41; Williams, 1989, p. 97; Rippon, 1997b, p. 216).

— THE USES OF RECLAIMED LAND —

Inevitably the uses to which reclaimed land was put varied from place to place and through time. Much that has been written refers to general trends in land use, but there are a series of detailed assessments, notably from the Somerset Levels and Romney Marsh, which act as pointers to specific agrarian regimes.

The enclosure of a piece of peat fen might constitute no more than an assertion of individual rights over turbary or common pasture (Hallam, 1965, p. 88; Williams, 1982, p. 103; Dinnin, 1997, p. 21); equally it might be the first stage in full scale drainage and reclamation leading to arable use, and this could occur relatively quickly, certainly within thirty years (Tatton-Brown, 1988, p. 106). This conversion of marshland to arable commenced at an early date: charred cereal grain from the Norfolk Marshland has revealed cultivation in mid-Saxon times (Murphy, 1993, p. 75).

Enclosure and limited drainage would convert fen to meadow and pasture, and the frequency with which meadow land is documented suggests that in many of the Lincolnshire marshland parishes, and probably in Cambridgeshire and Norfolk as well, new fenward reclamations initially provided pasture and hay crops (Hallam, 1965, p. 86). Similarly the reclamations from the peat fens also tended to provide pasture and meadow, as the records for Crowland Abbey's demesne reveal (Darby, 1940, p. 49; Hallam, 1965, p. 178), though the Thorney example cited above makes it clear that fen intakes would be cultivated in the right conditions. Much of the medieval embanking that occurred in Essex, in the Kent marshlands (Everitt, 1986, p. 61), and along the Lancashire coastline seems to have been geared to the creation of pasture (Wilkinson and Murphy, 1995, p. 207; Whitney, 1989; Hallam, 1965, p. 133; Cowell and Innes, 1994, p. 210), while the marshes of the Sussex Ouse were prized meadow by the fourteenth century (Brandon, 1971b, p. 96), and the embankments along the Derwent in Yorkshire allowed meadowland to be enriched by winter floods (Sheppard, 1966, p. 17). Many years ago Darby stressed that meadowland was a particularly important element in the Fenland economy (1940, p. 61) and this point appears to be equally valid for other wetland areas. The value of an acre of meadowland frequently exceeded that of arable: on Battle Abbey's estate at Barnhorne in 1305 meadowland was costed at 18d, marshland arable at 12d and upland arable at 6d (Brandon, 1971a, p. 69). Comparable figures have been produced for some of the Glastonbury Abbey clayland estates such as Berrow and Lympsham where around 1300 an acre of meadow could be 2s, twice the value of arable land (Rippon, 1997b, p. 223 based on Keil, 1964). In the Gwent Levels in 1291, Tintern's Moor Grange had pasture valued at double what could be raised from their other holdings (Rippon, 1996b, p. 79).

Nevertheless, figures from an early fourteenth-century register of extents for Glastonbury reveal that regardless of the relative values of meadowland and arable, the latter covered much larger areas, particularly on the higher, coastal claylands where between 70 and 80 per cent of the land was cultivated (Williams, 1970, Fig. 11). Analysis of nine thirteenth-century extents for the wapentake of Kirton reveal a similar pattern on the Lincolnshire siltlands (Hallam, 1965, p. 183). In the Gwent Levels it is assumed that there was a mixed farming regime, although there is little solid evidence on which to base this belief (Rippon, 1996a, p. 96). On the Pevensey Level the newer reclamations would be turned over to pasture while the older intakes would be cultivated (Brandon, 1974, p. 113; Dulley, 1966, p. 30). In the Fens the ubiquity of fragmented medieval pottery in the soil announces the widespread manuring of cultivated land (Hayes and Lane, 1992, p. 59; Silvester, 1988, p. 165), and recent fieldwork on Romney Marsh has built up a similar picture with the added refinement of some negative evidence for manuring on a Canterbury Cathedral Priory estate, which is known from documentary sources to have concentrated on livestock production throughout the Middle Ages (Reeves, 1995, p. 89).

No detailed analyses of broad trends in cropping systems has been attempted, most of the details being derived from specific estates at particular times. Inevitably perhaps, different patterns of arable land use are encountered in the reclamations in England and Wales, and Hallam put it succinctly in noting that 'the proportions of sown to mown varied greatly as need dictated' (1965, p. 175). The higher and drier Fenland silts grew wheat, but other estates belonging to Crowland Abbey produced barley, oats, rye and leguminous crops such as peas, beans and vetch (Hallam, 1965, p. 179). Barley, beans, oats and peas were the main crops around Sowy in the central Levels (Williams, 1970, p. 55); Canterbury Cathedral Priory's marshland estate of Ebony had on average three-quarters of its arable down to oats in the early fourteenth century (Smith, 1940, p. 33), but on some of their other estates at the end of the thirteenth century legumes reached 35 per cent of the arable (Gross and Butcher, 1995, p. 109).

— THE LATER MIDDLE AGES —

The decline in population resulting from the Black Death of the mid-fourteenth century, and the climatic downturn and perhaps rising sea levels from the later thirteenth century, had a marked effect on reclamations which were labour-intensive and appeared increasingly vulnerable to exceptional weather conditions (see Hallam, 1965, p. 132). Norfolk Marshland, for instance, was flooded twelve times between 1250 and 1350 (Darby, 1940, p. 57). There were also frequent complaints about the failure of landholders to fulfil their obligations to maintain and repair drainage works. This was the case with Wisbech Fen which a jury noted was drowned and valueless in 1439 (Darby, 1940, p. 165).

New reclamations are apparently rare, yet it would be unwise to generalise too much: there were some areas such as Walland Marsh and the Pevensey Levels where the intake of land continued in the fourteenth and fifteenth centuries (Gardiner, 1988, p. 115; Tatton-Brown, 1988, Fig. 9.1; Salzmann, 1910, p. 44); and though reclamation is considered to have ended in large parts of the Fens by the thirteenth century if not earlier, exceptions do occur, such as the new land taken from the peat fen of Billingborough in 1418 (Hayes and Lane, 1982, p. 24).

Broadly, it appears that attempts were made only to maintain the existing systems, sometimes by Commissions of Sewers, the first of which were appointed in the 1250s, sometimes by local lords or communities (Rippon, 1997a, p. 245). Whaplode Sea Bank was constructed as a result of the 1286 agreement mentioned above (Hallam, 1965, p. 136); in 1422 the Old Podike protecting the common of West Fen at the southern end of Norfolk Marshland was abandoned and replaced by the New Podike 3km to the south (Darby, 1983, p. 37); and at the end of the fifteenth century Bishop Morton of Ely built the eponymous Leam that diverted the waters of the Nene for nearly 22km on a straighter course towards the Wisbech outfall (Darby, 1940, p. 168).

Around the Severn Estuary Rippon has detected a phase of 'adjustment and reorganisation rather than desertion' with a move towards pastoralism (1997a, p. 17). The Wowwall near Weston-super-Mare was one of a number of new constructions made during the fourteenth and fifteenth centuries (Rippon, 1997a, p. 245).

On the Welsh side of the Severn, flooding during this period resulted in the abandonment of Goldcliff Priory (Rippon, 1997a, p. 242), and the inundation of most of Margam Abbey's coastal possessions (Williams, 1984, p. 278). The whole sea wall was abandoned in the fifteenth century and a new line adopted several hundred metres inland (Rippon, 1997a, p. 242). The reclaimed marshland of the Barnhorne estate on the Pevensey Levels required frequent repair works in the second half of the fourteenth century, but winter floods in 1409 left much of the land unusable during the following summer and further floods in the 1420s effectively ended cultivation, to be replaced by permanent pasture (Brandon, 1971a, p. 82). A similar disruption is suggested for other Sussex wetlands, such as the reclaimed estuary of the River Ouse (Brandon, 1971b, p. 97).

Elsewhere lower-lying fields may have been abandoned and this could be the context for the relict field systems that have been detected at Huntspill and Mark in Somerset (Rippon, 1997b, p. 210), in Holland Fen, Lincolnshire (Hallam, 1965, p. 70) and in Tilney Broad Fen (Silvester, 1988, p. 65).

— Conclusions —

The picture that emerges is one of continuous exploitation of marshland throughout southern Britain during the late Saxon and medieval centuries, frequently on a massive scale. Reclamation generated wealth, though perhaps not universally (Williams, 1982, p. 87), for the potential productivity of the intakes was generally greater than their upland equivalents. But it also necessitated constant maintenance and communal effort which was not always forthcoming. A combination of factors in the late Middle Ages served to demonstrate just how fragile the human hold on the wetlands was, and in some areas, such as the Gwent Levels and in the marshlands along the south coast, ground was lost rather more quickly than it had been won.

CHAPTER 10

Post-medieval drainage of marsh and fen

CHRISTOPHER TAYLOR

— INTRODUCTION —

By 1500 vast areas of marshland and fenland all over Britain had been reclaimed. Some of the former marshlands afforded rich arable land, and parts were occupied by farmsteads, hamlets and villages. Former tidal flats were protected by sea banks, behind which farmers became wealthy from the valuable pastures. Along the fen edges piecemeal but often extensive areas of reclaimed fields existed, surrounded by banks and divided by drainage ditches.

Yet despite this, large areas of fenland and marsh still lay in their natural state. They ranged from rough pasture, through oak and alder woodland, carr, sedge fen and reed swamp, to open water, as well as stretches of salt marsh, swept by every tide. They included large parts of the Somerset Levels, most of the deep peat fens of Lincolnshire and Cambridgeshire, much of the land around the Isle of Axholme in south-east Yorkshire and parts of the Norfolk Broads. In addition, the numerous peat-filled hollows and more extensive mosses of north-west England, in Shropshire, Cheshire, Lancashire and Cumbria, were also still mainly in their natural state, as were the carrs of Holderness and the Vale of York. Unreclaimed, too, were the edges of the estuaries of rivers such as the Thames and the Sussex Ouse and large areas around The Wash and along the east coast. Yet 400 years later almost all of this had been drained and reclaimed and much of it was amongst the most valuable agricultural land in Britain. How was this achieved and what made it possible?

— DEFINITIONS —

Three terms used in this chapter require definition. 'Drainage' is the creation of channels and embankments, the construction of sluices and the erection of water-lifting machines, all of which enable unwanted water to be removed from land. It is also the continuous process of water removal. 'Reclamation' is the creation of fields and the improvement of soils and vegetation in order to begin or increase agriculture by cultivation or grazing. These distinctions are important, for reclamation usually succeeded the initial drainage and was in turn often followed by a third development, 'settlement', or the establishment of farmsteads and hamlets on reclaimed land.

Reclamation could usually only take place if the initial drainage was perceived to be successful, and settlement only when reclamation was thought to be complete. These processes usually occurred in quick succession. Thus around the Isle of Axholme the land was first drained between 1626 and 1629, reclaimed in 1630 and at least partly settled by 1634 (Thomlinson, 1882). But if the process of drainage was unsuccessful, reclamation and settlement could only take place much later. Thus the North Holland Fens, north-west of Spalding, Lincolnshire, were 'drained' in the 1630s by the construction of the 38km long South Forty-Foot Drain. But because of the inadequacy and consequent abandonment of this drainage the fens were not properly reclaimed. Not until after 1765, when improvements were made not to the drainage but to the outfall in the river Witham, was reclamation begun and settlement established (Darby, 1966, pp. 48, 147).

— Drainage technology —

By the end of the medieval period most of the basic principles of the drainage of wetlands, evolved over centuries, were well known and widely used. These included the embanking of rivers, the construction of sluices, the use of counter-banks and drains and sea walls, catchwater drains and the cutting of diversion channels to increase the flow of water and to reduce silting. Tidal sluices or clysts, with self-acting gates which closed automatically on a rising tide or flood, were common. The only important piece of technology the medieval drainers are said to have lacked was a method of lifting water.

Yet it is possible, even likely, that such machines did exist by the fifteenth century. Windmills for lifting water were known in Holland before the middle of the fourteenth century and it would be surprising if these useful devices had not spread to Britain. The problem is one of terminology. The Latin *molenda* in medieval documents could mean water-mills, windmills or even horse-mills. Only the location of such mills might give a clue as to their function. Thus a 'mill dike' on the banks of the river Welland at Spalding, recorded in 1395, could well have been situated near a drainage-mill. Certainly drainage-windmills existed in the Cambridgeshire fens by the sixteenth century (Hills, 1967, p. 13).

None of the available techniques for drainage, known by the sixteenth century, changed very much until the nineteenth century. Drainage channels became larger and drainage schemes more complex and extensive, but these had little to do with advances in engineering or in drainage methods. Most ditches and channels, even the very largest, were still being dug by unskilled labour, using mainly hand tools, until well into the nineteenth century. Structures such as sluices and machinery such as windmills were all capable of being built by reasonably competent local blacksmiths and carpenters, perhaps with the aid of local masons or bricklayers. Even the techniques of land surveying needed for any large drainage schemes were mostly available by the seventeenth century (Tyacke, 1983). Both the history and the interest of post-medieval fenland and marshland drainage and reclamation lies, therefore, not just in the technology of drainage but in the interplay between it and a number of other very different factors. These must now be examined.

HYDROLOGY

Through the centuries the drainers had acquired considerable knowledge of the principles of water movement and related matters. Even so, hydrology was a science slow to develop and imperfectly understood, perhaps until the twentieth century. Certainly in the seventeenth century, for example, no one knew what caused the flooding of rivers. This led to interminable arguments as to how to solve the problem (Darby, 1966, pp. 35-7). Some people believed that flooding would be prevented and drainage assisted by putting rivers into straightened or re-cut deep channels where, safely between banks, they would flow with greater velocity and so remove water more quickly. Another school of thought believed that the problem lay with the outfalls, where the rivers reached the sea. To keep the channels there clean and scoured, either the volume of water or its speed had to be increased and incoming tides had to be shut out by sluices to prevent silting. Neither side was completely right, yet both methods and others were employed and still are. But there were always difficulties over which solution took priority in the inevitable situations where money for capital works was in short supply.

The problem of the shrinkage of the surface of land when water is removed was an equally difficult one to solve (Darby, 1966, pp. 104-13). All land compacts as water is removed. Thus the surface of waterlogged silt marshland or peat fenland will fall during the process of drainage. Over much of the reclaimed estuarine and coastal siltlands this lowering by anything up to a metre was a serious matter, but it was still usually possible to remove water by gravity at low tide. Even where the siltlands were eventually lower than the adjacent rivers and watercourses, drainage-windmills could scoop the water from one drain to another, provided the 'lift' was not too much. But in the peat fenlands of eastern England, in parts of the Somerset Levels, as well as elsewhere in northern England, it was a different matter. When water is removed from peat it shrinks rapidly, dries out, disintegrates following bacterial action and is blown away. The seventeenth-century drainage engineers, particularly the Dutch who had worked on the siltlands of Holland, had never seen the phenomenon of peat shrinkage and did not understand the process. They and their successors continued to believe that most drainage in the peat fens could be achieved by gravitational flow, despite evidence to the contrary. As late as 1809 even the great engineer John Rennie still believed that the Cambridgeshire fens could be drained without pumping (Hills, 1967, p. 7).

The reality was very different. As soon as drainage work in the fens of eastern England, on the Levels or around Axholme was begun the peat surface dropped. Within a very short time the ground surface of the fen fell well below that both of the main rivers and channels crossing the fens and of the internal drains. Two examples of this are instructive. In 1852 the great Whittlesey Mere, south of Peterborough, was drained. Within two years the surface of the adjacent fen had fallen by over 2.5m (Hills, 1967, pp. 7, 127-8, 143). And at Wedmoor Fen, also in Cambridgeshire, the embanked waters of the New Bedford river, cut into the underlying peat in 1651, now flows 7m above the land next to it (Figure 10.1). This phenomenon began to develop in the mid seventeenth century and has continued ever since. It was only coped with by the construction of thousands of drainage-windmills from the mid seventeenth century onwards, by steam engines from the early nineteenth century, and by diesel-driven and electric pumps in the twentieth century. The danger to the fens from

Figure 10.1 The Bedford rivers, Cambridgeshire. The greatest of the seventeenth-century drainage works, viewed here from the north-west, stretch for 30km across the fens. The left-hand channel is the Old Bedford river, cut in 1637 to shorten the circuitous Fenland course of the river Ouse by some 20km; the right-hand channel is the New Bedford river added in 1651. Between these lie The Washes, a vast reservoir which in winter and in times of flood holds the water from the upland rivers and prevents it inundating the adjacent fens. As a result of the lowering of the ground surface through the shrinkage of the peat these fenlands are now up to 7m below the level of the Bedford rivers. The diagonal channel across The Washes in the middle distance is the river Delph, an eighteenth-century cut intended to speed up the drainage of The Washes in spring and to allow them to be used for early pasture. (© Cambridge Committee for Aerial Photography.)

this situation was exacerbated in the early stages of drainage works. Before the nature of peat was understood, banks lining rivers and major watercourses were themselves often constructed of peat. These too shrank and then collapsed under the pressure of floodwater, usually with catastrophic results. Indeed, in some places banks continued to be built of peat well into the nineteenth century, although whether this was due to ignorance or economy is not clear (Taylor, 1981, p. 172). The remedy, rebuilding hundreds of kilometres of bank using clay laboriously dug by hand out of pits on the fen edge and then transported across the fens by barge, was expensive and time-consuming.

— ADMINISTRATION —

The success of drainage as well as the continuous maintenance then required was dependent on other factors vital in the history of fenland and marshland. These were the overall organisation and administration of drainage and the provision of revenue for both capital expenditure and maintenance. In medieval times most drainage was carried out in a piecemeal fashion by individual farmers and landowners, albeit often on an extensive scale. The problems that inevitably arose from the conflicts of interests involved, as well as the difficulties of raising money for the work, were partly solved by the introduction of Commissions or Courts of Sewers from the fourteenth century onwards. These organisations suffered the drawbacks of most similar medieval bodies. The Commissioners were local landlords with their own interests, corruption and tax evasion were usually widespread, and there were few powers of enforcement. Yet these Commissions survived in some places well into the post-medieval period, indeed in the Somerset Levels until the twentieth century. They thus hampered attempts at drainage improvement.

Elsewhere, by the sixteenth century, when the dissolution of the monasteries resulted in the advent of new landlords, events moved more quickly. The new owners were often successful merchants and lawyers who had made their way in a new world where reason and logic held sway and where science, influence and the law could be made to work to their advantage. As a result, ideas for comprehensive rather than piecemeal drainage schemes began to evolve, together with suggestions for the long overdue administrative systems which might achieve them. But such changes were not possible until the conclusion of what has been called the 'Tudor Revolution in Government'. Political necessity led to the legalisation of the Reformation by parliamentary statute rather than by royal despotism, placing both parliament and its decisions on a new footing. Thus was established the supremacy of statute on which the modern state still rests. For the new visionaries who aimed to drain the fens and the marshes, acts of parliament were the means by which their aims and actions could be put on a legal footing. As a result, from the 1560s onwards, acts of parliament were increasingly used to authorise drainage works at both local and national levels.

The most important of the early statutes was the General Drainage Act of 1600. This was significant in the history of drainage for it established a legal mechanism for financing drainage works. Until that time few landlords, small farmers, or even communities could find the capital for anything but small-scale drainage and reclamation. The 1600 Act sanctioned the granting of a proportion of any area of newly drained land to outsiders in return for their providing finance. So appeared the Adventurers, individuals who adventured their capital, and the Undertakers, who undertook to carry out drainage work. Many of the seventeenth-century and later drainage schemes were only possible because of this system. The great drainage works on the Isle of Axholme in the 1620s, involving the diversions of the rivers Don and Idle, and the even greater achievements in the Cambridgeshire fens of the 1630s and 1650s, which included the 30-km long New and Old Bedford rivers, were effected in this way (Figure 10.1). All were paid for by the granting of large areas of what became known as Adventurers' or Undertakers' Lands to outside financiers and contractors. Even small-scale drainage schemes were based on these principles. Between 1630 and 1646 attempts were made to drain The Fleet, a 12-km long lagoon behind Chesil Beach in Dorset. A group of financiers, called Adventurers, backed this far-fetched scheme which inevitably failed (Taylor, 1970, p. 134). Larger schemes, seemingly better thought out, also failed. One was the draining of King's Sedgemoor on the Levels. Both James I and Charles I, royal successors to the medieval monastic landlords, tried to organise outside finance by grants of land, but the attempt collapsed for reasons discussed below (Williams, 1970, pp. 95–100).

Later, the use of parliamentary acts for drainage matters became common. One of 1663 set up the Bedford Level Corporation which, with its powers of taxation and of administration of the Cambridgeshire fens, effectively organised the primary drainage there until the twentieth century. Other acts included one of 1669 which was another and equally unsuccessful attempt to drain King's Sedgemoor, and one of 1666 which authorised the Earl of Manchester and his Adventurers to drain Deeping Fen in Lincolnshire (Darby, 1966, pp. 78, 81; Williams, 1970, p. 112).

In the eighteenth century acts of parliament were used to establish another form of drainage administration. The old Commissions of Sewers and later bodies such as the Bedford Level Corporation or the Middle Level Commission (Act 1810) were responsible only for the principal drainage channels, rivers and sluices. The internal drainage of the fields themselves remained the responsibility of individual farmers and landowners. Particularly in the Cambridgeshire and Lincolnshire fens, where peat shrinkage was considerable, the need for drainage mills required more than the usual co-operation. A more organised system of water removal was needed. The solution was to use acts of parliament to authorise the establishment of Internal Drainage Commissions (later Internal Drainage Boards). These comprised locally elected landowners who were responsible for the establishment and maintenance of the drainage of compact areas of fen and marshland. They were empowered to levy taxes, borrow money, employ staff and construct drains and drainage mills. The first Commission was set up in 1727 at Haddenham, Cambridgeshire (Darby, 1966, p.119; Finney et al., 1995) and from the middle of the eighteenth century onwards many acts of parliament authorised the establishment of others.

Another form of parliamentary legislation used, this time for reclamation, were enclosure acts. From the seventeenth century onwards, but mainly in the eighteenth and nineteenth centuries, such acts were used to enclose common fields and wastes all over the country. But these acts were increasingly employed to enclose and reclaim open fenland and marsh. The first enclosure act for Somerset was for the reclamation of Common Moor, near Glastonbury, in 1722 (Williams, 1970, p. 112). It was on the Levels that these acts proved particularly useful. In the eighteenth century the old Commissions of Sewers were still responsible for general drainage but had decided that they had no authority to undertake new works. The enclosure acts got round this problem, and another that had developed, that of rewarding the Crown or its agents with up to one-third of all reclaimed land, by defining new, legally binding actions. Similar acts were also used to reclaim many of the mosses of north-west England.

From the eighteenth century onwards acts of parliament were also increasingly used to authorise and obtain finance for major engineering works aimed at improving drainage over wide areas. Thus in Cambridgeshire an act of 1795 sanctioned the construction of the Eau

Brink Cut, a 5-km long diversion of the river Ouse. Likewise the North Level Main Drain, a 10-kilometre long cut in south Lincolnshire, was authorised in 1831. In the Levels, an act of 1819 sanctioned the building of various channels and sluices, while in Axholme an 1814 act allowed the Drainage Commissioners to erect 'engines' and to carry out major works.

— SOCIAL FACTORS —

While technology, organisation and administration could enable drainage and reclamation to be achieved, it was necessary for the people involved to believe that the works could be effected and that they would be worthwhile. Often individuals or small groups refused to accept what was seen by others to be sensible or obvious. One result of this was the often considerable time-lag between drainage proposals and their implementation.

In addition, although drainage and reclamation was frequently assumed to be of benefit to major landowners and outside investors, it was a disaster for the smaller farmers and landless peasants. For undrained fen and marsh was the basis of a way of life that was luxurious by comparison with that of similar people elsewhere. Fens and marshes not only provided valuable grazing land but were often of a high enough quality to allow the bringing in of stock from elsewhere for fattening. Even more important, these lands gave even the poorest people access to fish, wildfowl, timber, peat, reeds, sedge and hay. Thus when fen and marshlands were drained, reclaimed, and parts given to outsiders, local people with common rights faced ruin (Thirsk, 1953a; Thirsk, 1953b).

On the completion of the drainage works in the Axholme fens in 1629, the 5000ha of commons at Epworth were reduced by grants to investors to under 2500ha. Small wonder then that commoners everywhere used every means at their disposal, both legal and illegal, to prevent drainage and reclamation. Yet despite court cases and riots the commoners eventually and inevitably lost and a way of life stretching back to pre-Roman times was destroyed. Sometimes the end was sudden. In June 1638 some 'forty or fifty men gathered in a fen called Whelpmoor... Common to Ely and Downham [Cambridgeshire]... to throw down the ditches which the drainers had made for enclosing their fen grounds from the Common'. The crowd was dispersed with violence, the ringleaders imprisoned and the reclamation continued (Darby, 1966, p. 61).

Elsewhere commoners were able to prolong opposition for decades. In the Somerset Levels attempts by the Crown to drain King's Sedgemoor began in 1618 but were finally abandoned in 1655. Further attempts were made between 1660 and 1688 but nothing came of them. The moor was not finally drained until the late eighteenth century. The reasons for this failure are complex, but the principal one was the opposition of the numerous commoners, many of whom were also freeholders. As such they elected not only the Commissioners of Sewers but also the local Justices of the Peace, often the same people. The commoners were thus able to exert pressure on what was effectively part of the county establishment (Barnes, 1961, p. 154). Another place where commoners fought for their rights over a long period was Otmoor in Oxfordshire. The drainage there was first mooted in 1787 but the arguments dragged on for years. An act authorising the work was passed in 1815 but its implementation was not attempted until 1829. At that stage riots, intimidation and even shootings occurred and troops were sent in. It was not until 1835 that the work was completed (Brown, 1967, pp. 34-52; Lobel, 1957, pp. 70-1).

The periods when drainage and reclamation occurred varied from area to area, often because of either opposition or delay in providing a suitable administrative body. As has already been seen, local Drainage Commissions appeared in Cambridgeshire in the early eighteenth century and were common by 1800. By contrast, in the Levels the cumbersome Commissions of Sewers continued to oversee drainage until well into the nineteenth century. Then the 1877 Somerset Drainage Act finally put internal drainage on a well organised footing.

Sometimes the difference in drainage history varied from parish to parish. In south-east Cambridgeshire for example, Bottisham, Swaffham and Burwell Fens formed a compact block of land which by the 1750s was drained by a multitude of drains and windmills, all maintained by individual farmers. In the 1760s attempts were made to form a Drainage Commission that would create and run a unified drainage scheme. But the landowners of Burwell refused to join. They believed that drainage would be expensive and would reduce their profits from grazing and peat digging. The Swaffham and Bottisham proprietors went on alone and in 1767 obtained an act to establish their own Drainage Commission. One of the numerous improvements the new Commission carried out was the construction of a counterbank to protect Swaffham Fen from the adjacent, and regularly flooded, Burwell Fen. It was not until 1840 that the Burwell landowners agreed to have a coherent drainage scheme and by then it was physically impossible to join the Swaffham one. A separate and totally unsatisfactory system was devised and as a result Burwell Fen was gradually abandoned. Only in 1940 was adequate drainage achieved (Taylor, 1981).

Another instance of fenland conservatism or inertia was the slow adoption of steam engines for drainage. Steam engines had been developed for pumping in the first half of the eighteenth century and by 1800 most deep mines in Britain were equipped with steam-driven pumps (Buchanan, 1982, p. 247). Yet steam engines were not used for fenland drainage until well on into the nineteenth century, the first being erected in 1816 at Sutton St Edmunds, Lincolnshire (Hills, 1967, p. 75). And it was not until 1830 that the first steam engine was built in the Levels (Williams, 1970, p. 160). By the mid nineteenth century they were in evidence on all fenlands but their slow acceptance has never been satisfactorily explained.

Diesel-driven pumps, when they arrived, were accepted more quickly. Although not developed until the 1890s, by 1913 diesel engines had been installed in a number of places and by the 1920s they were common. The speed of acceptance here was partly due to a change in attitude to technology, but also to an understanding of the increase in efficiency in pumps and in capital, maintenance and labour costs. The later introduction of automatic electric pumps after the Second World War also reduced costs, but their appearance was dependent on an integrated power system that was not really possible until the Rural Electrification Programme began in 1953 (Electricity Council, 1987, p. 67). And this, too, was the result of social changes and demands as much as of advances in technology.

The effect of social attitudes on drainage is nowhere better seen than in the huge engineering works of the 1950s and 1960s in the eastern Fens. The 40-km long Cut-Off Channel was designed to intercept all the flood water of the rivers Lark, Little Ouse and Wissey and carry it north to the 18-km long Flood Relief Channel between Denver and King's Lynn. But behind the practical engineering lay the perceived security that these works gave to the inhabitants of the Fens there, who had suffered particularly in the 1947 floods. They now expected, as of right, protection from such disasters.

— CLIMATE —

A factor often claimed as important in the history of drainage and reclamation is weather and climate. Storms, heavy rainfall and tidal surges are usually very obvious and the consequent floods figure largely in any history of wetlands. The slower and more subtle changes in climate over long periods are less obvious. Yet it is not easy to see what direct part either weather or climate played in the post-medieval history of the fens and marshlands. There are certainly great difficulties in correlating long-term climatic changes with periods of reclamation or abandonment of wetlands. The so-called Little Ice Age (1550–1700) does not seem to have had a direct influence on the amount, speed or timing of reclamation nor on the changing profitability of farming, except in a very indirect way (Wigley et al., 1981). In many places reclaimed land was often inundated, homes destroyed and lives lost. Yet by and large the long-term effects were limited, provided there was a will for, or profit in, recovery. The classic instance is the great 1947 flood in the eastern Fens and along the East Anglian coast, often said to have been the worst in history in its severity and extent (Darby, 1966, pp. 254–5). Yet most of the damage was repaired within a year and its effects, physical and economic, hardly visible in two. The real impact of climate is when it combines with other, more important, factors such as economic changes.

— Economics —

Probably the most important factor in the history of the drainage and reclamation of wetlands, already touched on obliquely, is the changing economic situation at national or international levels. Local economics, whether it was the cost of coal for steam engines, or the balance between the expense of embanking a marsh compared with the rarity of a tidal surge, were significant. But much more important were wider economic conditions. These have been ignored or misunderstood by both contemporaries and later local historians who have often seen more apparently obvious factors as significant. It has been periods of bad weather and floods leading to the abandonment of land and financial difficulties for farmers which were, and are, often assumed to be the problem when the real answer lay in the changing economic climate.

In all wetland environments there is a delicate balance between the success and the failure of agriculture. Flooding, which will always occur, is not the factor that controls prosperity. When agriculture is profitable it is worth draining and reclaiming land. At such times, even if there is flooding, it can be overcome and success ensured. And surplus profits can be ploughed back into expenditure on flood protection and drainage machinery which reduces the danger of future flooding. But at times when agriculture is only marginally profitable or losing money, capital expenditure is reduced, maintenance cut back and recovery from the inevitable flood damage is slow or non-existent. Thus if the history of the fens and marshlands is really to be understood it is the broad economic changes that have to be identified.

Around 1500 Britain, and indeed Europe, was emerging from the late-medieval economic depression. In the period 1500 to 1640, under the stimulus of a rising population, increasing prices and thus mounting land values, the demand for new land and the more efficient use of existing land became intense. Overall in this period prices rose by over six times and cereal prices by more (Bowden, 1985, pp. 1-117; Thirsk, 1953a; Thirsk, 1953b). These developments, together with the new landlords and new ideas noted earlier, provided the impetus for drainage and reclamation during this time. The great drainage schemes on the Isle of Axholme and in the Cambridgeshire fens, notably by the Dutch engineer Vermuyden, are the best known, largely because of the enormous engineering works associated with them. But smaller schemes in the Somerset Levels, the reclamation of the marshes along the Thames, around the Norfolk and Suffolk coasts, on the south coast in Hampshire and Sussex, as well as large areas of the Lincolnshire fens, were even more impressive in extent if visually less exciting. With the success of this drainage and reclamation went settlement on these new lands. Farmsteads were quickly established on the marshlands and even on the deep, unstable, peat fens. On the fens these farmsteads were often erected on the more stable silt beds and side banks of extinct streams and rivers.

After 1660, however, the tide began to turn. The huge additions to agricultural output, all over Britain as well as in the wetlands, began to slow down price rises and farming became less profitable. Reclamation continued, for example in the Lincolnshire fens, along the Humber estuary and on the Lancashire mosses (Allison, 1976, pp. 133-6; Darby, 1966, p. 148; Cowell and Innes, 1994; Hall et al., 1995, pp. 126-7; Middleton et al., 1995, pp. 75-6, 101-8), but the process stagnated. Other factors, such as peat shrinkage in the eastern Fens, opposition from commoners and the silting of the estuaries also played their part, and it was these as well as the regular flooding that were most obvious to contemporaries. Yet all these were symptoms of, rather than reasons for, the recession.

By 1750 the next economic upturn was under way with renewed agricultural prosperity following from industrialisation and urban growth. As a result, reclamation of wetlands gathered pace again and in the century to 1850 perhaps more land was taken in from fens and marshland than in all the previous centuries. The onset of the Napoleonic wars brought further price increases for agriculture and thus even more drainage and reclamation. Acts of parliament allowing enclosure and authorising Drainage Commissions and major engineering works were merely the administrative result.

What actually took place on the ground is hard to comprehend and difficult to explain without reducing it to a mere catalogue of events. Everywhere in the country reclamation and drainage went on apace. It occurred in the Vales of Pickering and York, in Holderness and even in the Humber estuary (Allison, 1976, pp. 167-8; Sheppard, 1958; Sheppard 1966; Figure 10.2). In north-west England the carrs of north Shropshire were all reclaimed between 1750 and 1850 while the great mosses of Lancashire disappeared in the forty years after 1800 (Cowell and Innes, 1994, pp. 162-4; Hall et al., 1995, pp. 126-7; Hey, 1984, pp. 65-6; Middleton et al., 1995, pp. 175-6). In the eastern fenlands vast areas of land were enclosed by acts of parliament in this period and in the first half of the nineteenth century the last of the great fen meres was drained and divided into fields (Darby, 1966, p. 201; Hills, 1967, p. 143). The extensive coastal marshes around The Wash, around the north Kent coast and elsewhere were also embanked and drained (Darby, 1966, p. 201). And alongside all this went improvements to existing drainage schemes with the construction of huge channels, the introduction of steam engines and the technical improvements of pumps. The arrival of centrifugal pumps after 1851, when the first was shown at the Great Exhibition (Heathcote, 1877, p. 12), was important in this respect.

By 1850 almost all the fens and marshlands in Britain had been reclaimed. It only remained to keep them in good heart. And until 1870, as elsewhere, wetland farming was extremely profitable. Yet this was only the calm before the onset of the storm of perhaps the greatest agricultural depression in British history. The reasons for this are beyond the scope of this book but they included the results of cheap imports of both grain and meat from the expanding lands of the New World, improved transportation methods, the introduction of refrigeration and the political ideology of free trade. This depression, with a brief respite during the First World War, lasted until the end of the 1930s (Mingay, 1994, pp. 194-244). Inevitably it hit wetland farming particularly hard when the cycle of falling income, reduced drainage maintenance and flood protection, floods, abandonment of

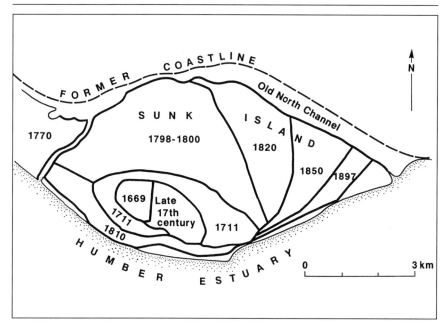

Figure 10.2 Sunk Island, Humber Estuary. Natural accretion of sandbanks along the north side of the estuary began in the sixteenth century and by 1560 a 'sunk' island was visible at low water. In 1669 it was large enough for some 8ha to be embanked and reclaimed and by the early eighteenth century another 600ha had been embanked, several farms established and a chapel built. Another 1,100ha of land, divided into six farms, were added between 1798 and 1800 and 150ha in 1810–11. During the following ninety years a further 900ha were reclaimed in a series of embanking progammes, adding five more farms. By 1900 Sunk Island was an island no more; it had become part of the mainland. Yet the surviving lines of the now largely redundant embankments still mark the progress of its reclamation. (Source: based on Allison, 1976, Fig. 19.)

land and even less income, began. Gradually arable fenland reverted to pasture, then to rough grazing and was often given up completely. In 1919 a report (Stewart, 1919) claimed that every part of the Somerset Levels had been subjected to flooding in 1916 when 30,000h of land were under water. In 1929 almost all the peat land in the Levels was flooded for six months (Williams, 1970, p. 223). Much the same situation prevailed in the eastern Fens. When Alan Bloom, better known for his gardening achievements, bought land in Burwell Fen, Cambridgeshire, in 1938 it was largely under water (Bloom, 1944, pp. 29–44).

There were some successes. The 1930 Land Drainage Act finally swept away the complex overlapping administrative systems for drainage that had developed since the fourteenth century by establishing new Catchment or River Boards. It also gave the Boards supremacy over the Internal Drainage Boards, thus helping to solve the age-old problem of parochial self-interest. Even so, by 1939 the future of agriculture in the fens and marshlands appeared bleak.

All this changed with the Second World War. The desperate need for more food meant that money was no object. The marginal wetlands were re-drained, improved and ploughed up regardless of cost. Subsidies and deficiency payments became the norm (Bloom, 1944; Mingay, 1994, pp. 245-69). Since 1945 the most important influence on the fens and marshlands has been the new social environment which has determined the policies and actions of engineers, economists and politicians for fifty years. The inhabitants of the wetlands have, like everyone else in Britain, come to expect access to electricity, transport, social services, cheap food, and, particularly, security from floods. Since 1947 serious inundations have been almost unheard of. Government grants, up to 85 per cent of the cost of new drainage schemes, have been available, subsidies for agricultural produce have been provided and electricity for pumping is now ubiquitous. If the last fifty years of wetland history has taught historians anything it is that technology and engineering are the least important factors in drainage. Social, economic and political matters are much more significant. And if looked at carefully, the history of fens and marshland shows that this has always been so.

CHAPTER 11

Water meadows: their form, operation and plant ecology

ROGER CUTTING AND IAN CUMMINGS

— INTRODUCTION: THE RISE OF GRASSLAND — IRRIGATION

Floated water meadows have been described as one of the greatest achievements of English agriculture (Kerridge, 1953a). It is estimated that some 40,500ha of water meadows were constructed between the early seventeenth and nineteenth centuries (Adkin, 1933). The overwhelming majority of floated meadows were in the chalklands of Wessex, but they were also created, albeit to a lesser extent, in other regions. There are examples in areas as far apart as Wales, East Anglia and the north of England (Kerridge, 1953a; Feltwell, 1992; Wade Martins and Williamson, 1994); they were even introduced on a limited scale into Scotland during the early nineteenth century (Orwin, 1929, p. 41).

The term 'water meadow', used strictly and correctly, refers to an area of land where the quality and quantity of herbage has been improved for agricultural purposes through deliberate hydrological management. Although descriptive literature relating to water meadows is extensive, there appears to be little empirical work concerning either their physical operation or their biological productivity. This chapter addresses these issues, and also describes the provisional results of some research recently carried out at the Britford Meadows, near Salisbury, Wiltshire, in order to discover the precise effects of irrigation on grassland. Their plant communities are also considered.

— THE BEGINNINGS OF MEADOW IRRIGATION —

The word 'meadow' originates from the Old English *maedwe*, related to *mawan*, 'to mow' (Rackham, 1986). The emphasis on mowing fields for fodder still forms an important distinction between meadows on the one hand, and grazing pastures on the other. In practice, at different times during the annual management cycle watered meadows would provide both pasture and (later in the spring or summer) a mown hay crop.

'Flood meadows' are simply areas of low-lying ground prone to natural flooding: hence they are almost universally to be found on alluvial soils. Flooding dresses the meadow with silt and nutrients, but flood meadows require low energy conditions in the associated watercourse: only as the river loses competence in flood is there a significant fall-out of entrained fine sediments. Once the benefits of flooding were recognised, the next stage was its manipulation or encouragement. In Britain possible evidence for the deliberate flooding of grassland has been identified as early as the Bronze Age, at Flag Fen in Cambridgeshire. Here, soil cores have revealed the presence of flood deposits at locations beyond the statistical norm for temperate rivers: given the pastoral nature of the economy practised by the people inhabiting the site it is assumed that these meadows were flooded to improve the grass (French, 1992). Benefits would be largely derived from the deposition of nutrient-rich alluvial sediment, although it is possible that, as early as this, the way in which a covering of water might serve to protect grass from frosts was recognised.

The most dramatic example of deliberate flooding comes from much later, in the post-medieval period: 'warping', flooding land to enrich the soils with transported silt, was carried out in parts of Yorkshire (Creyke, 1845) and on the Somerset Moors (Williams, 1970). A variety of evidence suggests, however, that natural flooding of low-lying meadows was in fact more widely encouraged, and to some extent controlled, in medieval and post-medieval England. What was sometimes termed 'floating upwards' involved making a dam across a river and ponding back the water behind (Kerridge, 1967). Although the meadows so treated might benefit from the deposition of silt, and to some extent from frost protection, the practice was fraught with dangers, for where grass is covered with standing stagnant water for long periods anaerobic and potentially toxic conditions develop, causing considerable damage to the sward (Chapter 2). To offset these problems, systems of watering were introduced, which could be carried out under the control of the farmer. Although first documented at the start of the seventeenth century, these may have had medieval precedents; either way, it is such intervention – which involves the constant movement of water across the surface of the grass sward – which defines a true water meadow.

— FORMS OF WATER MEADOW —

– Catchwork meadows –

Catchwork irrigation systems were employed from at least the early seventeenth century on valley sides where streams or spring waters could be diverted into a 'flood dyke' running along the contours. This would overflow, the water flowing into further channels, and ultimately returning to a tail drain or directly to the parent stream. The objective was to provide an even flow of water across the field, and most historians agree that the main effects were to warm the sward, protect it from frost, and flush it with oxygen and any nutrients present in the water.

One example of a catchwork system is that at Clipstone Park Farm, Clipstone, Nottinghamshire, described in some detail by Denison (1840). Here 120ha of land described as having 'very little value' was improved, albeit with varying degrees of success. Here the difference in levels between the flood dyke and the river was approximately 17.6m and the best inclination of fall was reported to be 1 in 9 (Figure 11.1). The Clipstone system is also an excellent example of the extent of land preparation required prior to the water being 'thrown over'. Gorse and heather needed to be cut and burnt, the land slope levelled, ploughed and fallowed. Even after such rigorous preparations, the first release of water through the system revealed areas either too steep, or too shallow, to be successfully irrigated; as well as the unforeseen loss of water to abandoned rabbit warrens. Eventually, the working meadows allowed Denison to comment that 'the produce of the meadow is very great, exceeding all anticipation'.

The catchwork systems developed on Exmoor during the first half of the nineteenth century also formed an integral part of a land reclamation and agricultural improvement scheme (see Chapter 12). These probably represent the most extensive example of catchwork irrigation ever undertaken in England. Here, the hill slopes tend to be steeper than at most other sites, typically between 10 and 18 degrees (Cook, 1994). The horizontal gutters tended to be slightly inclined downslope in order to allow more efficient flow: a gradient of 1:396 has been quoted (Smith, 1851).

Catchwork systems were best suited to narrow valleys which contained

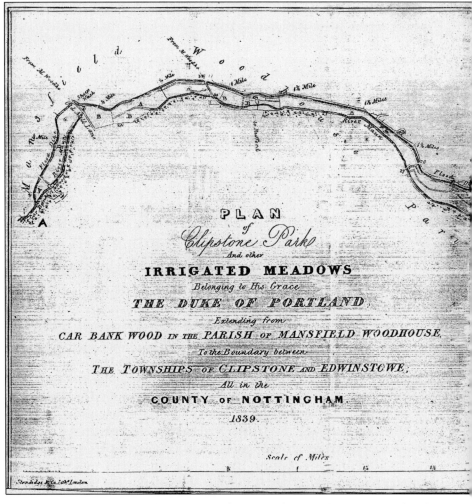

little or no floodplain alluvium, and the quality of the water utilised within them depended upon local soil and geological conditions. Because of the kinds of topography in which they occurred, catchwork systems were likely to use non-calcareous waters which were found to be less effective in stimulating the growth of grass than those used in the more familiar 'bedwork systems'.

– *Bedwork meadows* –

Bedwork systems, which were widely adopted in southern England in the course of the seventeenth century, were of far more significance than catchworks. They involved the diversion of water from a river via a sluice or main hatch into a channel known as the 'main carrier' or

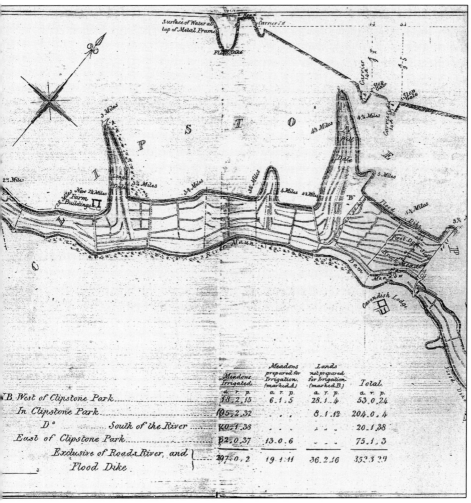

Figure 11.1 The 'catchwork' water meadows at Clipstone Park, Nottinghamshire, as illustrated by John Denison in 1840. (Source: Denison, 1840.)

'carriage'. This would carry the water to points of release which were generally controlled by a series of smaller hatches. The water, once released, would move along small, spade-cut channels known as 'floats' which were cut into the apex of a number of gently sloping parallel ridges, known as 'beds' or 'panes'. The rate and direction of this flow was controlled by placing turves, known as 'stops', in the channels. The water would be allowed to over-top the floats and flow down the sides of the ridges, to collect into peripheral drains; these then transferred the water to a larger tail drain, by which it was returned to the river, some way below the point of abstraction. The characteristic arrangements of drains and ridges, which can still be seen in many valley flood plains and river terraces in southern England, are often referred to as 'bedworks'. Figure 11.2 shows an aerial view of the Britford Meadows, Wiltshire, which are of this type; Figure 12.1 (next chapter) shows a bedwork system during floating.

The length of the ridges (and therefore the distance the abstracted water would need to be carried back) would be determined by the gradient from the river. This was seldom great in bedwork systems and was usually achieved by careful grading of the alluvial material. The ridges would be approximately 10-13m wide, with tapering floats to lead water along the apex, thus ensuring an equal flow of water across the length of the pane. These would be cut using a special spade (known as a 'coulter') with two vertical blades on either side of the main blade, and their width would reduce from approximately 0.5m to 0.3m. The difference in height between the float and drain was between 0.45m and 0.6m, each pane typically being some 4-6m wide (Atwood, 1964). The drains would also taper, from 0.45m at their maximum distance to 0.6m at the junction with the main tail drain. Once again this may have been to promote flow along the drain.

The cost of constructing water meadows was considerable (see Chapter 12). Not only was the construction of leats and drains expensive, but the subsoils within the area occupied by the meadow were often broken up to some considerable depth, in order to improve drainage. Where this was the case the turves would have to be removed prior to digging and then replaced. Some meadows took two or even three years to build (Atwood, 1964; Denison, 1840). In general, at least in those areas of southern England in which the practice was most widespread, this considerable investment was considered worthwhile, a clear indication that such systems brought real increases in productivity.

The time and effort often invested in the treatment of the subsoil indicates the importance placed on freely draining soils; the speed with which the water needed to be taken on and, more importantly, off the field is regularly stressed in contemporary literature. The geological context of bedwork systems is therefore crucial. Most appear to be associated with chalk rivers in southern England, whose valleys generally contain extensive spreads of Pleistocene gravels (IGS/SWA, 1979) which provide good subsoil drainage for infiltrating waters during floating. The storage capacity of the chalk aquifer provides a reliable flow in the main waterways; aquifer storage dampens the peak flows and sustains base-flow in rivers when compared with impermeable catchments, where flooding presents greater problems (Ward and Robinson, 1990, p. 241).

— MANAGEMENT OF WATERMEADOWS —

As already noted, floating was carried out in two distinct periods of the year; during the winter, and again during the late spring. Flowing water across the meadows during the winter was thought to insulate the soils sufficiently to protect them from frost and thus allow the early germination and growth of the grass; temperate grasses generally grow at, or above 5°C. Forcing an 'early bite' in this way helped alleviate the shortage of winter fodder to which farming systems were prone, and thus allowed larger numbers of livestock to be kept.

It was the job of the 'drowner', 'waterman' or 'meadman' to keep the water flowing evenly over the surface of the ground. The objective was to maintain a steady flow of water, 25mm deep, moving between the blades of grass yet not flattening them. Once the hatches and ditches were operational, and the water was turned onto the field, a 'steady state' flow could be established for between four and six days, gradually decreasing in duration between November and February (Sheail, 1971). A constant flow had to be maintained: standing water was considered (correctly) to be 'poisonous' to grass and was indicated by an accumulated 'scum' left behind when floating had ceased (Kerridge, 1967, p. 259).

Irrigation was most extensively employed within 'sheep-corn' husbandry systems, that is, within forms of cereal farming adapted to thin, nutrient-poor soils which depended on large sheep flocks for their fertility. The sheep were grazed by day on upland pastures, and folded by night on the arable, when fallow, after harvest or even when first sown. By increasing the numbers of sheep that could be kept through the winter (when the grass did not grow to any significant extent) the amounts of manure applied to the arable could be raised, thus enhancing crop yields; and the numbers of lambs being fattened for market could also be increased.

The stock were normally kept on the meadows until May, when they were removed and irrigation recommenced. This second irrigation was intended not to provide grass for direct consumption, but rather an enhanced hay crop – leading, once again, to an increase in the number of livestock which could be kept over winter. Floating in late spring would have the effect of eliminating any soil water deficit that had developed. Indeed, in some cases floating allowed a second hay crop to be taken in late summer. Claims have even been made that a third crop of hay was forced from the meadows, although there is little documented evidence for this (Sheail, 1971).

In the late summer and autumn stock, often including cattle, were returned to the meadows, thus maximising their use. During October, however, it was usual to turn them off again, so as to allow time to carry out repairs to carriers, drains and hatches in anticipation of the highly prized November flood waters.

The decline of water-meadow management during the later nineteenth and twentieth centuries was due to a number of changes in agricultural technology and practice. The introduction of artificial fertilisers, and the development of new forms of winter fodder (root crops, and ultimately oil cake) gradually undermined the need for such management, and the post-1870s agricultural depression, and twentieth-century labour shortages, hastened their disappearance. Today it is estimated that only 3 per cent of watermeadows remain operational (RSNC, 1991); many of those are Sites of Special Scientific Interest (SSSI) and some are managed for conservation purposes. The Britford Meadows shown in Figure 11.3 are thus designated. The Environmentally Sensitive Area Scheme (ESA), and Countryside Stewardship both now recognise the restoration of meadows as worthy of grant aid.

Research is urgently needed in order to enhance the management of the remaining examples of irrigated meadows. Precisely how floating served to promote an early growth of grass has been debated. Most contemporaries seem to have been aware that the insulating effect of the river water encouraged early germination and growth; but more general increases in productivity have, by some authorities, been ascribed to the flush of nutrients from both water and depositional sediment, and to the beneficial effects of dissolved oxygen.

The degree to which sediment flushes were effective has been questioned (Atwood, 1964); but eighteenth and nineteenth-century references to the particular benefits of 'dirty water' have been cited as evidence that this was a major factor in enhancing fertility (e.g. Atwood, 1964; Cowan, 1982; Feltwell, 1992). Indeed, during the middle decades of the nineteenth century the use of sewage, pumped on to flood meadows and water meadows, was advocated as a means of alleviating the (considerable) urban waste water disposal problem of the time (Sheail, 1996). Goodland (1958) has, in addition, argued that silt deposition from chalk streams was an ideal way of adding nutrients to the soil without the problem of acidification, because of the buffering effect of the dissolved calcium.

Doherty (1985) has shown that irrigated water meadows also featured a particularly abundant sward. Indeed, the winter flooding of these meadows may have suppressed the growth of perennial weeds due to the early germination and growth of the grasses, which in consequence out-competed them.

Different writers have thus suggested a number of possible ways in which irrigation may have enhanced grassland productivity. Yet, as Cowan (1982) has stated, 'the subject is well but repetitively covered in general terms but lacks published work on specific examples'. Most published work has looked at the flora of water meadows (Brian, 1991) but has not adequately approached the impact of management from the hydrological, temperature, nutrient and soil standpoint, despite past assumptions regarding the importance of these factors. What follows is a preliminary report on experimental work currently underway at one surviving example of a floated meadow, at Britford in Wiltshire, which is being carried out in order to supply the kind of experimental information needed if the practice of floating is to be fully understood.

— THE BRITFORD MEADOWS —

The Britford Meadows SSSI (SU 166274) covers 18.2ha and is one of the few 'bedwork' watermeadows still actively managed in England. The water is supplied from the adjacent Avon Navigation, a canal effectively acting as a meander cut-off on the Wiltshire Avon, approximately 4km south of the city of Salisbury. Assuming that the meadow is contemporary with the Navigation, it must have been constructed in the period 1675–1730; however, given the somewhat haphazard appearance of the pattern of floats and drains (Figure 11.2), it is likely that construction was piecemeal, with each field being developed separately. Since first established, the site has probably been subjected to uninterrupted management.

– Soils and soil water –

The soils of this section of the Avon Valley are mapped by the Soil Survey as falling within the Frome association, being typically a grey mottled silty clay loam, above calcareous gravel lying at a depth of

some 28–30cm. The soil water regime may vary locally, but in general prolonged waterlogging from ground water, and periodic flooding, gives a Wetness Class of IV or V (Findley et al., 1984). Being ancient pasture, topsoils exhibit a typical 'crumb' structure which imparts a high porosity to the surface horizons. One limited area within the SSSI is occupied by more peaty soils, of the Adventurers' association.

– *Soil water content profiles* –

Figure 11.4 shows the mean percentage soil water contents recorded

Figure 11.2 Aerial view of part of the Britford Meadows, Wiltshire. (© Cambridge University Committee for Aerial Photography.)

at five locations: six replicates were determined down-profile at each site. Four sites were efficiently watered, and one in the non-flooded area acted as a control. Subsequent readings taken fortnightly throughout the floating periods November 1995–February 1996 and November 1996–February 1997 were similar to the initial readings, suggesting a constant hydrological environment (Figure 11.4). Curves

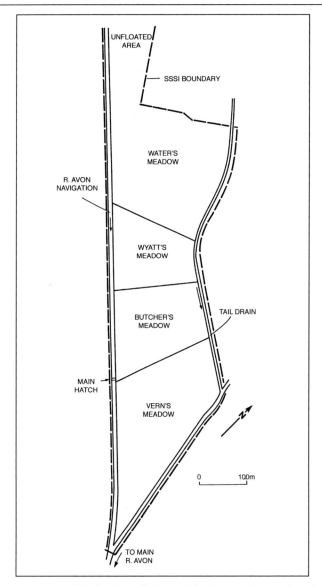

Figure 11.3 Generalised plan of the Britford Meadows, showing areas referred to in the text.

displaying a reduction in water content at depths of 6–10cm are interpreted as being due to the loss of crumb structure (and hence loss of macroporosity) beneath the surface 'A' horizon. The unfloated control site mostly maintained a drier profile for the upper 14cm than the floated sites.

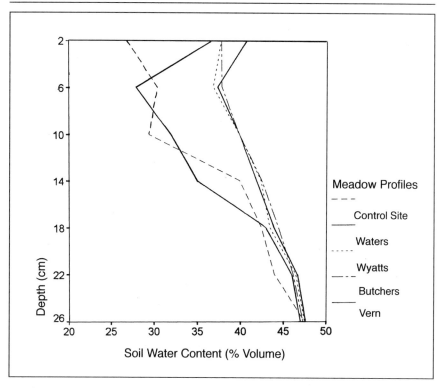

Figure 11.4 Mean soil water content (by % volume) in Britford Meadows experimental plots during floating period December–February (1995–6).

– *Infiltration rates and saturated hydraulic conductivity* –

The infiltration of the soil was measured in order to ascertain the rate at which the profile would reach saturation following the onset of floating, using a standard double ring infiltrometer. Typical initial rates were 30–40mm/minute, falling to 10–20mm/minute after about 10 minutes. For the small area of Adventurer's Series soils within the study area initial infiltration rates were 100mm/minute; these humose soils would, however, have required considerable amounts of water to remain effectively floated over the winter period and it is probable that they were never in fact managed in this way. During floating, steady state infiltration into the mineral soils was much lower, typically 0.5–7mm/hour.

High initial infiltration rates suggest rapid saturation but it is also important to rid subsoils of penetrating water in order to prevent anaerobicity. The saturated hydraulic conductivity of the gravel substrate is therefore an important influence on the transmission of soil water. The shallow gravels, often with a sandy loam matrix, underlying Britford Meadow permitted the auger-hole method (Van Beers, 1963) to be employed to measure horizontal saturated hydraulic conductivity (permeability). The values obtained were typically around 20mm/day. Using these data, the hydraulic conductivity may be classified as 'slow' (Thomasson, 1975) in absolute terms, yet the permeability is sufficient for effective operation of the meadow.

– Sediment load –

As has already been noted, early writers often expressed a preference for sediment laden water, or 'dirty water' as it was generally termed. Indeed, there are references to the reduction in the efficiency of the technique of floating across the water meadows, due to the sediment being deposited near to the point of abstraction, rather than evenly distributed across the grazed area.

Water samples were therefore collected across the Britford system to determine the sediment load. Sample points included the Avon Navigation, drains and carriers across the meadows, and various points on the tail drain. Table 11.1 shows mean monthly suspended solids on three sample dates.

TABLE 11.1 Mean monthly suspended solids collected from sample points on the Britford Meadows

	Mean Monthly Suspended Solids (mg/l)[1]		Mean Monthly Suspended Solids (mg/l)[1]	
Sample date	Avon Navigation	Tail drain	Carrier	Drain
November 1996	2.67 (0.05)	2.82 (0.17)	–	–
December 1996	2.63 (0.23)	2.69 (0.12)	3.85	3.93
January 1997	3.21 (0.70)	2.77 (0.14)	4.4	4.25

1. Standard deviation in brackets
n = 3 samples per month

These results imply that very little sediment is, in fact, brought on to the meadow from the Navigation, and there is little change in the overall sediment budget across the meadow. Indeed, the amount of suspended sediment was marginally higher in the drains and tail drain than in the Navigation, implying perhaps the entrainment and removal of fine particulates from the soil during floating, probably derived from drain clearance operations and cattle disturbance throughout the year.

These results may, in part, be explained by the particular character of the water being extracted from the Avon Navigation; the slow flow of the canal probably allowing the fall-out of the suspended load prior to abstraction by the main carriers. These results may thus be peculiar to the Britford site, although it would appear the water supply for floated meadows in southern England tends to be from chalk streams, which in general would be expected to exhibit low turbidity.

At Britford the main hatches are normally drawn for a few minutes prior to floating. This allows a considerable movement of water along the Navigation, and causes significant entrainment of sediment. The main hatches are then closed and those on the carriers opened. Each section of the meadow has been observed to take an average of 17 minutes to become flooded, and approximately 3 hours for the soil water profile to equilibrate, i.e. to become saturated. The initial flood of water across the meadow would be more sediment-charged than later flows; within 2 hours, however, the water appears to be clear in both the drains and carriers. It is probable that the initial opening of the carrier hatches allows for a pulse of sediment on each occasion, and this pulse is then available for deposition in the grass sward. After this initial floating, however, little further sediment appears to be added.

– *Dissolved load* –

The water flowing within a river system comprises a complex solution of dissolved compounds. A useful generic measure of dissolved solids is conductivity, and the water samples collected from the identified sampling points were tested using a conductivity meter (Table 11.2).

TABLE 11.2 Mean monthly conductivity values for water samples collected from sample points on the Britford Meadows.

Sample date	Mean monthly EC ($\mu S\ cm^{-1}$)		Mean monthly EC ($\mu S\ cm^{-1}$)	
	Avon Navigation	Tail drain	Carrier	Drain
November 1996	390	334	–	–
December 1996	410	430	440	460
January 1997	280	250	440	420
February 1997	340	350	380	395

n = 3 samples per month
Source: Environment Agency

Compared to the annual mean values of such catchments as the river Ouse (845 µS cm^{-1}) or the river Thames (553 µS cm^{-1}) these monthly values appear low (Shaw, 1983). However, the data only apply to the winter floating season, during which time greater dilution may be expected. Furthermore, the catchment is predominantly non-industrial and although higher values for dissolved salts may be expected from the surface drains of Salisbury, the outfall for the sewage treatment plant for this large urban area is downstream from the sample site.

Water flowing down the panes did not normally exceed 25mm depth and hence was generally within the grass sward. The contact with the atmosphere would be expected to maintain good oxygen levels in the irrigation waters, and the colder the water, the greater the maximum content of oxygen it can dissolve. Surface measurements across the field throughout the winter were typically above 90 per cent of saturation (i.e. within 10 per cent of the maximum oxygen content of the water for any given temperature). For comparison, water in stagnant pools and in shallow groundwaters at unfloated sites was typically below 60 per cent.

Nitrogen present in the system may be in the form of ammonia (NH_3), nitrite (NO_2) or nitrate (NO_3). In grazed grassland, ammonia will originate from animal dung and urine. In aerobic soils, this is free to oxidise to nitrite and nitrate, although the latter also derives from artificial nitrogen fertilisers.

Table 11.3 Mean monthly figures for the determination of dissolved nitrogen (NH_3–N, NO_2–N, NO_3–N) from samples collected on Britford Meadows

Sample date	Avon Navigation			Tail drain		
	NH_3–N (mg/l)	NO_2–N (mg/l)	NO_3–N (mg/l)	NH_3–N (mg/l)	NO_2–N (mg/l)	NO_3–N (mg/l)
December 1996	<0.03	0.014	6.387	<0.03	0.015	6.488
January 1997	<0.03	0.013	6.103	<0.03	0.013	6.136
February 1997	<0.03	0.015	6.442	0.05	0.015	6.877

n = 4 samples per month
Source: Environment Agency.

Table 11.3 shows the initial results of water analysis and appears to indicate a moderate level of nitrogen being brought on to the meadow system from further up the catchment. For comparison, the EU limit for NO_3–N in drinking water is 11.3mg/l. These results are low compared to some regions, but the accumulated input of nitrogen may lead to modest accumulations in the meadow in the course of the winter period.

– Microclimates –

Many earlier writers on the operation of watermeadows realised that water kept the soils relatively warm during the winter months. This, in turn, was seen as a cause of the early growth of the grass once the water had been removed. Monitoring of soil, air and water temperatures at the Britford Meadows between November 1996 and February 1997 (Figure 11.5) reveals that the temperature of the floated soils rarely fell below 0°C, and was often at or above 5°C, the point at which temperate grasses can grow. Air temperatures, by contrast, frequently fell below both thresholds during the period. Evidently, floating a meadow dampens the effects of extreme variations in the temperatures of the air above it during winter. This finding strikingly confirms the views of early agricultural writers, and of many modern historians, that one of the main benefits of winter floating came from its beneficial effects upon the ground temperature.

– THE INFLUENCE OF FLOATING ON PLANT – COMMUNITIES AT BRITFORD SSSI

We have seen how the principal aim of floating is to ensure a continuous flow of water across each meadow. This is achieved via carriers and drains, uniquely arranged to suit each meadow, in order to avoid or reduce accumulations of stagnant water that might impair the quality and yield of the grass. Two meadows were selected for study: Butcher's Meadow has long been floated, and is still managed in this way today; Water's Meadow is no longer so managed, and is effectively a flood meadow (Figure 11.3). In each meadow, random co-ordinates were used to position eight transects, four in each meadow, each laid out across the side of a separate pane. Each transect comprised three $0.25 m^2$ quadrats located at the top of the pane ('upper'); midway between the top of the pane and the drain channel ('middle'); and in the drain at the bottom of the pane ('lower'). Species-relative abundance estimates (converted to the Domin scale) were employed to obtain National Vegetation Classifications (NVC) using Tablefit (Hill, 1993). The character of the soil in each of the transects was also noted.

The dominant NVC community (Table 11.4) present in both the floated (Butcher's) and the flooded meadows (Water's) is MG11,

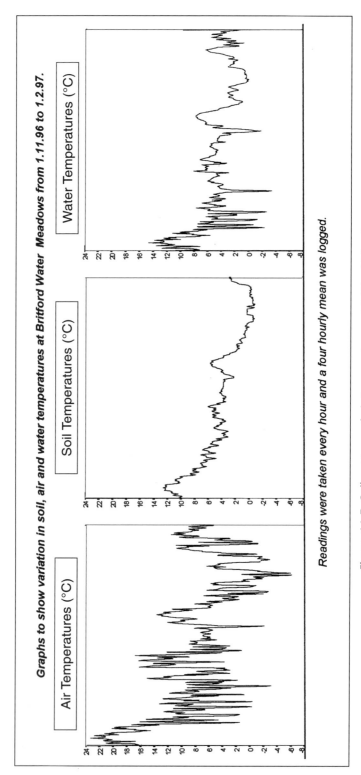

Figure 11.5 Soil, water and air temperatures at Britford Meadow.

Festuca rubra-Agrostis stolonifera-Potentilla anserina grassland (Rodwell, 1992; nomenclenture follows Stace, 1992). This community is characteristic of moist but free-draining soils. Notably, this vegetation type is generally recognised as being species-poor and this is confirmed by observations at Britford. In 1888, Fream had recorded a surprisingly high figure of eighty-five different species of flowering plants growing on the watermeadows in North Charford, bordering the Christchurch Avon in South Hampshire. However, a closer examination of Fream's data (Tansley, 1949, p. 572), suggests that he may have over-estimated the number of species by including those associated with the ditches. For example, Fream lists (*inter alia*), *Glyceria maxima*, *Phalaris arundinacea*, *Epilobium hirsutum* and *Veronica beccabunga*, all of which were, at Britford, confined to the drains and carriers. Sheail (1971) has suggested that floated meadows were relatively poor in species as a result of selective management by marshmen for the most useful plants. The relatively species-poor swards observed on Butcher's (and to a lesser extent on Water's) Meadows are probably likewise the result of weeding to promote the *Gramineae*, coupled with intense grazing, but to some extent they must also be the consequence of floating itself, as Doherty (1985) suggested.

It is true that, in general, a close similarity exists between the two meadows despite differences in management. Floating of Butcher's Meadow for up to three months each year does not appear to have altered substantially the plant communities when compared to Water's Meadow, which is flooded only when the level of the Avon Navigation rises. However, two of the communities identified, MG8 (in Water's Meadow) and S5a (in Butcher's Meadow), are notable anomalies. MG8 is a *Cynosurus cristatus-Caltha palustris* grassland (Rodwell, 1992), and is relatively species-rich (compared with MG11). This community contains fewer grasses and rather more dicotyledons such as *Caltha palustris*, *Filipendula ulmaria*, *Ranunculus repens* and *Ranunculus acris*. Its occurrence suggests that a gradual transition from a floated community (MG11) to a flooded community (MG8) may be occurring. Furthermore, mottling was more frequently observed in soil profiles sampled in Water's Meadow, subsoil conditions which would tend to support the notion that flooding, and resulting anaerobic conditions, may be gradually changing the plant communities.

On the other hand, the occurrence of S5a (a *Glyceria maxima* swamp community; Rodwell, 1995) in the lower pane in Butcher's Meadow suggests that floating might also occasionally result in the accumulation of stagnant water in certain areas (although management aims to reduce their occurrence). Furthermore, the lesser degree of mottling observed in the soil profiles of Butcher's Meadow suggests that groundwater influences are relatively weak. The main occurrence of the S5a community observed was in the ditches draining Butcher's Meadow. One ditch supported S14c, a *Sparganium erectum* swamp community, including *Nasturtium officinale*, *Apium nodiflorum* and *Typha angustifolia*. By far the most species-rich areas of the ditches and carriers (indeed, in the whole of Britford Meadows), occurred where grazing has resulted in open patches amongst the tall fen plants, enabling other species to colonise (c.f. Tubbs, 1978).

TABLE 11.4 Britford Meadows, Wiltshire. A comparison of the NVC (National Vegetation Classification) plant communities recorded on upper, middle and lower positions of the panes in Water's (flooded) and Butcher's (floated) Meadows and in the drains surrounding Butcher's Meadow.

Water's (flooded) Meadow (replicate)	Sampling location on pane	NVC types	Butcher's (floated) Meadow (replicate)	NVC types	Drains in Butcher's Meadow (replicate)	NVC type
1	Upper	MG9a	1	MG11	1	S5a
1	Middle	MG11	1	MG11	2	S5a
1	Lower	MG11	1	n.d.	3	S14c
2	Upper	MG11	2	MG11	4	S5a
2	Middle	MG8	2	MG11		
2	Lower	MG11	2	S5a		
3	Upper	MG11	3	MG11		
3	Middle	MG11	3	MG11		
3	Lower	MG11	3	MG11		
4	Upper	MG11	4	MG11		
4	Middle	MG11	4	MG11		
4	Lower	MG11	4	MG11		

N.d. = not defined; data analysed using Tablefit (Hill, 1993).

In the main, Butcher's and Water's Meadows have very similar NVC communities – notably species-poor, high-yielding MG11 grassy swards. Abandonment of floating may be resulting in a change to relatively more species-rich communities (MG8) as observations on Water's Meadow seem to indicate. However, by far the most species-rich communities were observed in the drains and carriers (S5a), especially where grazing occurred.

As already noted, the soils within the area studied are mostly mapped as the Frome Association (Findley et al. 1984): they are dominated by silty clay loam and clay loam profiles above gravels with clay loam, sandy clay loam and silty clay loam, typically at a depth of 30-70cm. Soil textures are therefore broadly similar in the two meadows, but the occurrence of mottling below about 30cm is more abundant in Water's (flooded) than in Butcher's (floated) Meadows, suggesting greater shallow groundwater influences in the former. The occurrence of *Junci* in the lower sections of the panes as an indicator of waterlogging must be treated with caution here as the germination of *Juncus* seeds is also encouraged by frequent disturbance (Moore and Burr, 1948) by, for example, cattle hooves.

In summary, this study of the vegetation communities at Britford Meadows has shown:

> Floating of Butcher's Meadow has helped to produce a grass sward that is characteristically species-poor (i.e. MG11).
>
> Abandonment of floating (for several decades) under a grazing regime appears to have resulted in an increase in species richness, the sward developing from MG11 (*Festuca rubra-Agrostis stolonifera-Potentilla anserina* grassland) to MG8 (*Cynosurus cristatus-Caltha palustris* grassland).
>
> The botanical richness attributed to watermeadows appears to lie predominantly in the drains and carriers rather than in the meadows themselves.

— CONCLUSIONS —

The research at Britford is still in its early stages but the data collected so far appear to suggest that floating had three main effects on low-lying grassland. Firstly, it served to maintain ground temperature during the winter at a slightly higher, and also more even, level than that of the air above. Secondly, the high proportion of dissolved oxygen within the water served to permit respiration within the soil biomass. Lastly, it may have decreased the extent of species diversity, encouraging the growth of grasses and discouraging a variety of 'weeds' of less importance as animal feed. The evidence collected so far suggests that irrigation did not increase the deposition of silt, or dissolved nutrients, to any significant extent.

Watermeadows are not only important from the point of view of agricultural history, or landscape archaeology (Cowan, 1982). They also represent a sustainable system of grassland and hydrological management. Today, only a handful of actively floated meadows survive, usually through the interest and enthusiasm of local land owners and farmers, yet some of these sites represent perhaps 300 years of uninterrupted land management. The skills of the 'drowners' or 'meadmen' have all but disappeared, and with them perhaps has gone a closer appreciation and understanding of the practical operation by which these meadow systems operate and produce grass for grazing and fodder. It is hoped that further work at Britford will continue to elucidate the detailed operational mechanisms.

As a whole, the Britford Meadows support a rich community of flowering plants and simultaneously produce high yields of grass without a single penny being spent on expensive fertilisers or herbicides. This must be a system of agri-environmental management highly deserving of attention from farmers and conservationists alike.

CHAPTER 12

The development of water meadows in the southern counties

JOSEPH BETTEY

— INTRODUCTION: THE SPREAD OF A NEW — AGRICULTURAL TECHNOLOGY

It was in the valleys of the chalk downlands in southern England that the practice of watering meadows through complex systems of hatches, sluices, channels and drains reached its fullest extent and became, for some three centuries after 1600, a vital element of the farming regime. The clear, fast-flowing chalk streams of Wiltshire, Dorset, Berkshire, Hampshire and Sussex, with their constant temperature, valuable nutrients and calcareous nature were ideally suited for watering meadows and for encouraging an early and abundant growth of grasses (see Chapter 11). From the chalklands the idea spread into neighbouring areas through the use of hillside or 'catchwork' meadows and to alluvial sites on non-chalk rivers. The remains of many thousands of acres of (now disused) water meadows survive along most of the chalkland valleys as witness to the former importance of this remarkable farming method. This chapter will describe the origins of the system in the Wessex downlands, the manner in which the practice spread so rapidly, the costs and difficulties involved in creating efficient meadows, the profits and benefits to be derived from them, and the essential part they fulfilled in the farming system of the region.

The earliest references to full-scale and complex watering of riverside meadows by artificial means, as opposed to the use of naturally-occurring winter floods, are found early in the seventeenth century. The idea of creating meadows that could be watered at will through a system of hatches and channels was publicised by Rowland Vaughan in his book *The Most Approved and Long Experienced Water Workes*, which was published in 1610 and described Vaughan's work in the Golden Valley,

Herefordshire. The Vaughans had family and business connections with the Herberts, Earls of Pembroke, whose widespread Wiltshire lands included many chalkland manors, and Vaughan wrote a dedicatory letter to William, Earl of Pembroke (Wood, 1897).

Although there is no direct evidence, it seems likely that it was through this family connection that water meadows were first introduced into the southern chalklands. The heavy initial expense of creating a water meadow meant that, at first, it was essentially a manorial rather than an individual enterprise, and the earliest references are to be found in manorial records. The first reference to a fully-developed system occurs in the manorial Court Book of Affpuddle, Dorset, where the landlord, Sir Edward Lawrence, was keenly interested in agricultural improvement. In 1605 there are references to ditches and channels being constructed in the meadows along the river Piddle, and in 1607 and 1608 complaints were made about the use of the water and disturbance to the ancient water course. In 1610, the year of publication of Vaughan's book, it was agreed at the Affpuddle manorial court that three men should be appointed to oversee the watering of the meadows, that the tenants should pay for the work in proportion to their holdings of meadowland, and that no one should interrupt the work or interfere with the channels (DRO, D/FRA/M1).

The active encouragement and interest of the lord of the manor, as well as pressure from the wealthier freehold and leasehold tenants in persuading copyholders to embrace the new ideas, is also evident at Puddletown, Dorset. At the manorial court held in October 1629 the lord, Henry Hastings, was present when 'a greate debate beinge theare had and questions moved by some of the tenants', it was agreed that one of the leaseholders, Richard Russell, and others should be allowed to continue with the work already started 'for the wateringe and Improvinge of their groundes in Broadmoor'. In this new and untried project, success was not assured, and the agreement contained a provision that 'yf yt shall appeare after the maine watercourse shalbe made throughe the said grounde thatt Improvement cannott be made upon some good part of Mr Woolfries groundes out of the same watercourse, ... thatt then the said Mr Russell shall fill in the said watercourse againe att his owne costes all alonge in Mr Woolfries grounde'. Richard Russell was evidently prepared to accept the risk of failure of at least part of the scheme in his enthusiasm to press ahead with it. (DRO, D/PUD/H2, fols 5-7, 8-9v, 17-22v, 60-61v, p. 189; Bettey, 1977). Clearly, the advantages of water meadows in providing early grass were quickly apparent to neighbouring manors along the

The development of water meadows in the southern counties

Figure 12.1 A drowner at work, Charlton-all-Saints, in the Avon Valley near Salisbury. The photograph was taken in February 1935. (© Rural History Centre, Reading University.)

Piddle and Frome rivers, and in his *Survey of Dorset* around 1630 the local landowner, Thomas Gerard, wrote of the river Frome passing 'amongst most pleasant Meadows, manie of which of late yeares have been by Industrie soe made of barren Bogges' (Gerard, 1980 edn).

Other Dorset landowners were also active in promoting the creation of water meadows. At Warmwell on a tributary of the river Frome a water meadow was constructed during the 1630s at the instigation of the lord of the manor, John Trenchard, an active land speculator and money lender. Theophilus, Earl of Suffolk, also encouraged the tenants on his Lulworth estate in agricultural improvement, and during the 1630s water meadows were laid out along the Frome at Burton, West Lulworth, Winfrith Newburgh and Bindon (Bettey, 1977, pp. 39-40). At Winfrith Newburgh in 1636 the Earl entered into an elaborate agreement with his tenants to share the costs 'in consideration the said Earle is pleased by way of watering... to improve the Meadow called Winfrith Mead'. The agreement was signed by the Earl and twenty-two tenants (DRO, D/WLC/E130).

Meanwhile, in south Wiltshire similar developments were taking place under the influence of the Earl of Pembroke. There are references to meadows called 'water mead', 'water close', and 'wett mead' in the court rolls of the Earl's manors of Chalke, Chilmark and Netherhampton during the 1620s, and in 1632 a detailed agreement for the creation of a water meadow was included in the court roll of Wylye. It is clear from this that the practice of watering was already well understood, and the agreement provided for the management of the meadow, the rights of the tenants, the necessary construction costs and payments for maintenance and management. John Knight of nearby Stockton was employed to 'drawe a sufficient and competent quantitie of water of the River of Wylye, out of the same River, sufficiently to water and flott all the said groundes or soe much thereof as by industry and art may be watered or flotted' (Kerridge, 1953b, pp. 38-40; Kerridge 1953c, pp. 105-18). John Aubrey, who had himself farmed at Chalke, recalled around 1680 that 'the improvement of watering meadows began at Wylye about 1635, at which time we began to use them at Chalke' (Aubrey, 1847 edn., p. 104). Aubrey also recalled the making of water meadows around Salisbury and along the Kennet in north-east Wiltshire at the same time. By the later decades of the seventeenth century, water meadows were already a well-established feature of Wiltshire chalkland manors, especially along the several rivers which converge on Salisbury (Kerridge, 1953b, pp. 112-15).

During the 1640s and 1650s there are numerous references to water meadows being constructed in Hampshire along the Test, Itchen and Meon, and in 1669 John Worlidge of Petersfield, Hampshire, could describe the watering of meadows as 'one of the most universal and advantageous improvements in England within these few years' (Worlidge, 1669; Bowie, 1987). Likewise in a report on the husbandry of Dorset made to the Georgical Committee of the Royal Society in 1665 by Robert Seymer of Hanford near Blandford Forum, Dorset, the water meadows of the chalklands were described as an established feature of farming practice 'the greatest improvement they have for their ground is by winter watering of it, if it lye convenient for a River or lesser streame to run over it' (Royal Society, CP, 10/3/10).

Further to the east, and outside the chalk downland areas, water meadows were in use along the Wey between Haslemere and Frensham by 1680, and aqueducts, sluices and channels survive in many places along the valley (Bowles, 1988, pp. 1-9).

— Problems of construction —

During the later seventeenth century water meadows continued to be constructed and extended throughout the chalklands. Expertise in their operation grew, and ever more complex systems were devised. For example, when a water meadow was constructed along the river Stour at Charlton Marshall, Dorset in 1659, 'able and sufficient carpenters' were obtained from Tolpuddle for constructing the hatches, and Henry Phelps of Turners Puddle, 'a Known Ancient, Able and Experienced waterman' was sent for to supervise the whole project, 'soe ordering the water whereby that the said groundes might be well watered... as farr as the strength of the River would cover'. The scheme was encouraged by the landlord, Sir Ralph Bankes of Kingston Lacy, and was supported by the Provost and Fellows of Eton College who had freehold rights in the meadows. A rate was levied on the occupiers of the land to pay for the work, and thirty-six tenants contributed £62 12s 6d (PRO, C5/58/15). The work of laying out a meadow, with precise levels and the essential network of weirs, hatches, channels and drains, was inevitably complex and laborious. It required considerable expertise to divert a fast-flowing chalk stream and ensure an adequate covering of the meadow. The detailed estimates for renewing one of the hatches on the Wyndham estate at Upavon, Wiltshire during the mid eighteenth century provide an indication of some of the work involved. Piles had to be driven into the bed of the river Avon, secure foundations laid, and a false riverbed constructed, with two walls of solid ashlar each 14.6m long and 2.4m high on either side to contain the river. The heavy wooden hatches had to be strong enough to withstand the pressure of the water; they were fixed to the foundations by iron clamps and embedded in the walls (WRO, 2667/22).

An elaborate scheme for watering the meadows on the Avon south of Salisbury was executed during the 1670s by John and Leonard Snow, the stewards of Sir Joseph Ashe and later of his widow, Lady Mary Ashe. Joseph Ashe possessed the manor of Downton and much land along the valley of the Avon, although he lived at Twickenham. The scheme involved a long main carriage to bring the water from Alderbury some 4km upstream. This supplied water to various Ashe properties before reaching Downton, and the construction involved no less than twenty-one agreements with landowners, farmers and millers, the construction of several bridges, as well as a weir and hatches on the river Avon. The eventual cost was more than £2,000, more than double the original estimate made by John and Leonard Snow. They justified the expense by pointing out that 194 acres (79ha) of meadow could be watered and that the value of the land had been greatly increased:

> 70 acres which were worth £80 per annum are now worth £180;
> 50 acres which were worth £20 per annum are now worth £100;
> 74 acres which were worth £74 per annum are now worth £148.
> (WRO, 490/891-904; 1946/H21, H25. Pugh, 1968, pp. 54-5, 71-7).

In order to convince the tenants of Sir Joseph Ashe that the elaborate work required to create water meadows would be to their advantage, the stewards produced a list of arguments in 1676. These may be summarised as follows:

1. There would be a great increase in the crops of hay.
2. Men could keep more sheep and cattle and thus their arable ground would be improved.
3. There would be an increase in corn and grass for fattening cattle and for butter and cheese.
4. Taking water for the meadows need not hinder the miller.
5. Disturbance or damage caused to other mens' ground by digging carriages, channels etc. would be properly compensated (WRO, 490/890).

Further evidence of the rapid spread and potential profitability of water meadows during the late seventeenth and throughout the eighteenth century is provided by the numerous legal disputes between manorial tenants and millers over rights to water, and the agreements drawn up to resolve these. At Tarrant Rushton, Dorset in 1646 the miller complained that his watercourse was '*obstructat et divertit causam innundationi pratori*'; the construction of water meadows on the river Ebble at Odstock, Wiltshire could only proceed after long negotiation with the owner of Odstock Mill and agreement as to the dates and times of watering and when the hatches should be shut or drawn (Salisbury MSS, Hatfield House, 9/3; WRO, 490/900). At Nunton near Salisbury in 1698, the tenants agreed to pay Lord Coleraine £3 per annum for the use of water from the stream above his paper mill. At Longparish on the Test in Hampshire five farmers each paid the miller 2s per acre per annum for the use of his water on three days a week and also agreed to have their corn ground at his mill (WRO, 490/894; Bowie, 1987, p. 157).

The development of water meadows in the southern counties

By the early eighteenth century the practice of watering meadows had spread throughout the downland valleys of Wiltshire, Dorset and Hampshire, and along the Lambourn and Kennet in Berkshire, the Lavant, Rother and Arun in west Sussex and the Wey in Hampshire and Surrey (Young, 1793, p. 44; Short, 1985, pp. 270-313; Wordie, 1985, pp. 329, 331). In 1771 Arthur Young praised the 'exceedingly rich watered meadows' along the Kennet from Marlborough to Hungerford, and 'the uncommon importance of having a command of water to throw at pleasure over grasslands'. While travelling through Dorset he also noted the improvements made by watering, and described how one man had risen from day labourer to tenant farmer through his expertise and labour in 'bringing water over all the land that he possibly could' (Young, 1771, III, p. 31, IV, pp. 306-20). This rapid spread leaves no doubt of the benefits and profitability of water meadows, and of their importance in the sheep/corn husbandry of the chalk downlands. Expert observers could describe the meadows of south Wiltshire as 'almost indispensable' and the value of the meadows as 'almost incalculable' and that 'the early vegetation produced by flooding is of such consequence to the Dorsetshire farmer that without it their present system of managing sheep would be about annihilated' (Davis, 1794; Claridge, 1793). In the Hampshire chalklands it was regarded as essential for a successful farm to have water meadow at one end and downland grazing at the other (Vancouver, 1810, p. 267).

—THE OPERATION OF THE— WESSEX WATER MEADOWS

As noted in the previous chapter, the primary purpose of the Wessex water meadows was to provide early grass for the sheep flocks which were such an essential feature of the farming system of the region. A secondary purpose was to provide an abundant and reliable hay crop to sustain the flocks during the winter. The fertility of the thin chalkland soils could only be maintained by the dung of the sheep folded intensively on the land. It was from wheat and barley that chalkland farmers derived most of their income, and large sheep flocks were essential for the production of satisfactory crops. The water meadows provided a lush growth of grass during the hungry weeks of the early spring, in March and April, when the hay stocks were exhausted and some six weeks before natural growth had started. Thus they breached the age-old barrier to agricultural progress, and by enabling farmers to keep larger sheep flocks provided the means to extend arable acreages and increase yields of corn.

As usual, William Cobbett described the situation succinctly. Of the Wiltshire downland in 1826, he wrote:

> *Sheep* is one of the great things here; and sheep, in a country like this, must be kept in *flocks*, to be of any profit. The sheep principally manure the land. This is only to be done by *folding*; and to fold you must have a *flock*. Every farm has its portion of down, arable, and meadow; and, in many places, the latter are watered meadows, which is a great resource where sheep are kept in flocks; because these meadows furnish grass for the suckling ewes early in the spring; and indeed, because they always have food in them for sheep and cattle of all sorts. (Cobbett, 1912 edn., II, p. 40.)

In most chalkland manors the operation of the water meadows was closely controlled by the 'waterman', 'meadman' or 'drowner' who imposed a strictly regulated calendar. By Michaelmas each year the hatches, channels and drains would have been checked and cleared. Watering could then begin for short periods, depending on the weather and on the availability of water which was often subject to agreements with millers or other landowners. Under the watchful eye of the waterman a Wessex meadow could produce a lush growth of grass by mid March, and this could be used to feed the ewes and lambs. These were allowed to feed in controlled sections behind hurdles for short periods each day, 400 or more couples to the acre, and straight from the fresh new grass were then folded upon the land destined for the spring-sown barley. This was the period when the water meadows really repaid the trouble and expense of their construction and maintenance. In 1794 Thomas Davis, who was the experienced steward on the Longleat estates in Wiltshire, wrote that:

> The water meadows of Wiltshire and the neighbouring counties are a branch of husbandry that can never be too highly recommended... none but those who have seen this kind of husbandry can form a just idea of the value of the fold of a flock of ewes and lambs, coming immediately with bellies full of young quick grass from a good water-meadow, and particularly how much it will increase the quantity and quality of a crop of barley. (Davis, 1794, pp. 35-8.)

The sheep were removed from the water meadows at the beginning of

May, because at this time they were liable to contract liver fluke and foot-rot from the damp pasture. By now the meadows would have been eaten bare, and the natural downland grazing would be available for the sheep. The meadows could then be watered for a few days and thereafter left for hay. The hay had to be cut young, about mid June, or it rapidly became coarse, but yields of two tons to the acre were common. (Young 1808, p. 222; Vancouver 1810, pp. 276-7; Bowie 1987, p. 153). Occasionally the meadows were watered again to produce a second hay crop, but generally they were used for grazing cattle during the later summer when drier conditions meant that they would not 'poach' the ground or damage the intricate network of channels and drains.

Between them, the water meadows and the sheep flocks were the pivot around which downland farms revolved, being the requisites to the production of the cash crops of wheat and barley. By the nineteenth century the meadows were so much part of the landscape that in 1878 Thomas Hardy could describe the view from the height of 'Egdon Heath' across 'meadows watered on a plan so rectangular that on a fine day they look like silver gridirons' (Hardy, 1968 edn, p. 205).

— COSTS AND PROFITS —

The costs of creating a water meadow and the expenses of operating it varied enormously, depending upon the location, gradient, water flow and other demands upon the water, as well as the original condition of the meadow, so that precise figures can be misleading. The water meadow at Marsh Moores, Twyford, Hampshire cost approximately £3 6s per acre to construct in 1670-2, and Otterbourne Mead, Hampshire just over £6 per acre in 1730-1 (Bowie, 1987, p. 155). Writing of Dorset in the 1780s George Boswell estimated the average cost as £6, while William Stevenson in 1812 put the cost at £7-£8 per acre (Boswell, 1779, pp. 108-9; Stevenson, 1812, p. 370). Thomas Davis, describing the Wiltshire meadows in 1794, put the cost at between £12 and £20 per acre, but stated that watered land increased threefold in value (Davis, 1794, 30-8); in the early nineteenth century in Dorset and Hampshire the cost was estimated at £8 or £9 per acre (Moon and Green, 1940). It was generally agreed that once constructed, the annual costs were low.

The authors of the various county surveys or *General Views* produced for the newly-created Board of Agriculture during the 1790s and early

1800s were enthusiastic advocates of the benefits of water meadows, and gave figures to support their arguments. William Mavor described the 56 acres (23ha) of water meadows laid out along the Thames on the estate of Edward Lovedon at Buscot. These cost just over £2 per acre and the annual expense was less than 10s per acre, while the work 'has doubled the original value and produce of the land' (Mavor, 1809, pp. 369-70).

Charles Vancouver provided figures for the profit to be made from an acre of water meadow in Hampshire. He calculated that the expense per annum, including interest on the capital invested in making the meadow plus labour, maintenance, rates and tithe, would amount to £5 18s 6d per acre. The income included spring feed for 400 ewes and lambs at 7d per couple per week, totalling £1 13s 4d; a crop of hay worth £4 17s 6d; a second hay crop or rent of the late summer grazing £2 12s 6d; giving a figure of £9 3s 4d, to which could be added the value of the sheep fold straight from the meadow which he calculated was worth at least 16s 8d per acre, giving (he calculated) an overall total income of £10, or an annual profit per acre of £4 1s 6d (Vancouver, 1810, pp. 276-7). In the *General View* for Sussex the Revd Arthur Young produced a similar calculation showing an annual profit per acre of £4 8s 6d exclusive of the value of the sheep fold (Young, 1808, p. 222).

Further evidence of the profitability of water meadows is provided by the high rents which could be charged for them. At Compton, Hampshire in 1738 the rent of unwatered meadow was 20s per acre while water meadow was let at 40s (Thirsk, 1985a, p. 68). Unwatered meadows on the Itchen around Winchester in 1808 were said to be worth only a third as much as the water meadows (Bowie, 1987, p. 157). At Britford, Wiltshire the water meadows were let in 1839 for 80s per acre, while dry meadow was rented for 40s (Bowie, 1987, p. 156).

Thomas Davis summed up the situation in Wiltshire when he wrote of the water meadows that 'the improvement in the value of the land is astonishing' and that their usefulness for the sheep flocks 'is almost beyond computation'. They were 'so useful as to be indispensable in south Wiltshire' (Davis, 1794, pp. 34-5). The consequence of this profitability was that throughout the chalk downlands water meadows were pushed to the limits of the valleys. By the end of the eighteenth century there were said to be 6,000 acres (2,430ha) of water meadow in Dorset and 15-20,000 acres (6-8,000ha) in south Wiltshire (Stevenson, 1812, pp. 304-5; Davis, 1794, p. 34).

— Publications and the — Spread of Information

In spite of the widespread use and profitability of water meadows, it was not until the later eighteenth century that detailed descriptions of their construction, layout and management were published. Rowland Vaughan had given precise details of his experiments in 1610, but thereafter agricultural writers such as Blith (1653: MacDonald, 1908) and Worlidge (1669) confined themselves to praise of the early grass and productivity obtained by watering. One of the first practical treatises on the subject was written by a Dorset farmer, George Boswell (1735-1815). Boswell was born in Norfolk, but came to Dorset as a young man and spent the rest of his life in the district around Puddletown, near Dorchester, with only occasional visits to his relatives in Norfolk. It is likely that he came to Dorset in the service of Robert Walpole, second Earl of Orford, who had acquired the manor of Puddletown through his marriage in 1724 to Margaret, daughter of Samuel Rolle of Heanton, Devon. Later, Boswell engaged in farming and other business enterprises on his own account, becoming a tithe collector, maltster, surveyor and mercer. His main interest remained in agriculture and in the spread of improved methods and ideas, especially the techniques and benefits of watering meadows. Throughout his life he was a practical farmer, renting land on the Earl of Orford's estate and later on the estate of a local land-owner, James Frampton of Moreton. Frampton obviously thought highly of Boswell, and in 1790 granted him a lease of Waddock Farm at Affpuddle, one of the best farms on his estate, with fine new buildings, enlarged farm house and extensive water meadows along the river Frome. In his Memorandum Book Frampton recorded the lease and noted that he had charged a low rent:

> ... keeping rather below than above the market price, and with great allowance always and indulgence to active, spirited servants, particularly to such an uncommon one as Mr Boswell, whence the Landlord has many advantages, especially a choice of sensible, active, responsible and good neighbours. (James and Bettey, 1993, pp. 119-20.)

Boswell contributed to the influential *Journal* of the Bath and West Agricultural Society which was founded in 1777, but his national reputation was established in 1779 with the publication of his book *A Treatise on Watering Meadows*. This provided a full account of the methods and benefits of watering meadows, with detailed descriptions and diagrams of weirs, hatches, channels, drains and trenching tools. Boswell's book was essentially a practical guide and, for example, gives much advice on the diversion of streams, surveying of channels, and the siting of hatches. The construction of sluice gates is described in detail with numerous diagrams and suggestions. Boswell stressed his own practical knowledge and experience, 'not the effusion of a gareteer's brain, nor a Bookseller's job, but the result of several years experience' (Boswell, 1779, pp. 1-2). The book achieved considerable success, and a second, larger edition was published in 1790. It brought Boswell to the attention of contemporary agriculturalists including Robert Bakewell. It also brought him into contact with George Culley, a leading farmer from Northumberland. A series of letters from Boswell to Culley, covering the years 1787 to 1805, provides much detail about Dorset farming, livestock management, social conditions and especially about sheep and water meadows (James and Bettey, 1993).

Contemporary with Boswell was the Revd Thomas Wright, who had served as a curate at South Cerney near Cirencester, Gloucestershire during the 1780s. He had been greatly impressed by the water meadows along the river Churn, a small tributary of the Thames, and in 1789 he published *An Account of the Advantages and Method of Watering Meadows by Art* (Wright, 1789). Later Wright became rector of Ould in Northamptonshire and in 1790 published a second edition of his book, adding *as Practised in the County of Gloucestershire* to the title. As a result of Wright's enthusiasm water meadows were developed along the Churn at Down Ampney and along the nearby river Coln. These were described by Thomas Rudge in 1807 who recommended their wider use in the county (Rudge, 1807). George Boswell was much less impressed by the Revd Wright's practical knowledge of the subject and wrote scathingly of 'a new treatise upon Water Meadows by a Clergyman in Gloucestershire. People in their zeal for a Cause they've espoused generally think they can never point out the advantages too strongly – but by over doing it often hurt it' (James and Bettey, 1993, p. 138). Nineteenth-century advocates included William Smith, whose *Observations on the Utility, Form and Management of Water Meadows* was published in Norwich in 1806.

— CATCHWORK MEADOWS —

The success of water meadows in the chalkland valleys encouraged farmers in neighbouring areas to experiment with catchwork or hillside

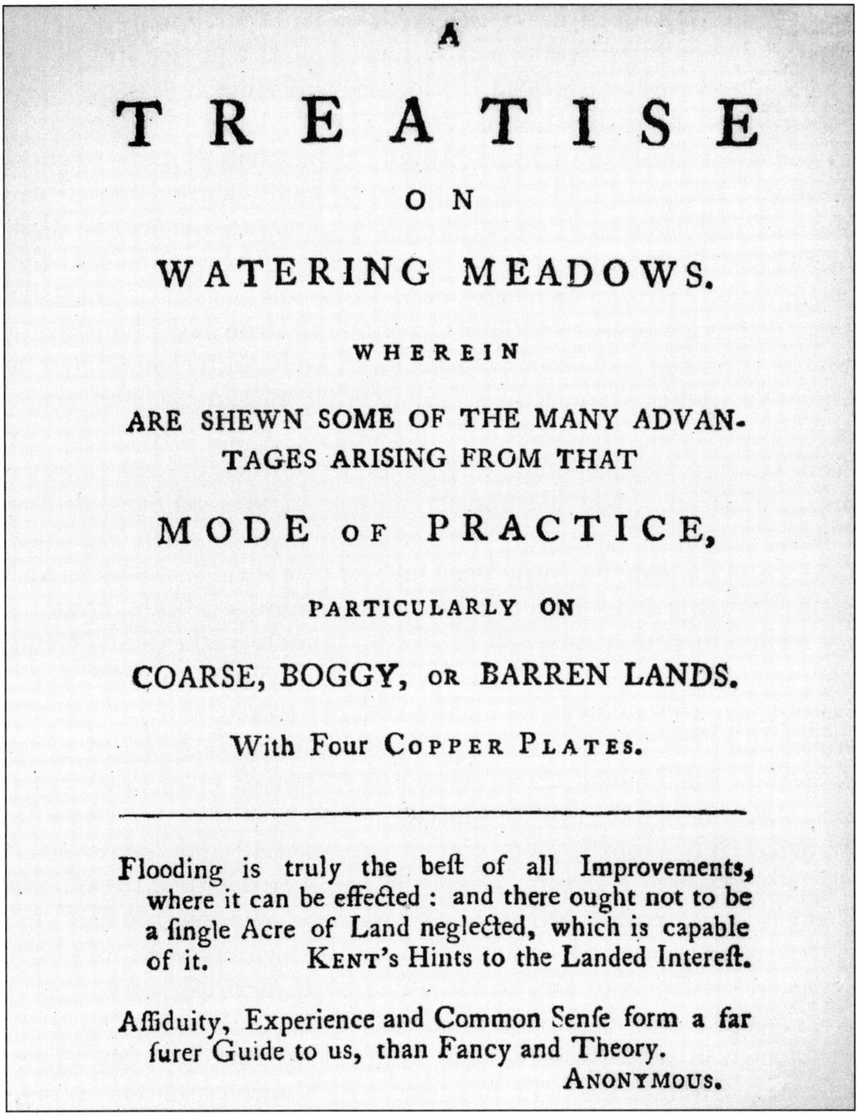

> A
> # TREATISE
> ON
> ## WATERING MEADOWS.
> WHEREIN
> ARE SHEWN SOME OF THE MANY ADVAN-
> TAGES ARISING FROM THAT
> ### MODE of PRACTICE,
> PARTICULARLY ON
> ### COARSE, BOGGY, or BARREN LANDS.
> With Four Copper Plates.
>
> ---
>
> Flooding is truly the beſt of all Improvements, where it can be effected : and there ought not to be a ſingle Acre of Land neglected, which is capable of it.   Kent's Hints to the Landed Intereſt.
>
> Aſſiduity, Experience and Common Senſe form a far ſurer Guide to us, than Fancy and Theory.
> Anonymous.

Figure 12.2 The frontispiece from George Boswell's A Treatise on Watering Meadows, *published in 1779.*

meadows. Unlike the 'floated' meadows of the chalk, catchwork meadows depended upon water brought in a leat along the contours of a hillside and made to overflow down to subsidiary 'gutters' or leats lower down the slope which in turn overflowed on the next part of the hillside. Although catchwork meadows were never quite so satisfactory or so intensively used as the floated meadows, they were cheaper to construct and easier to manage, and relic features of the hatches, leats and gutters can still be seen on many hillsides in west Somerset, west Dorset, and east Devon.

Catchwork meadows were also used on the chalk hillsides around Ashbury, Berkshire and along the escarpment of the Vale of the White Horse (Mavor, 1809, p. 371). The extensive remains of these hillside meadows can be seen in Dorset at Wynford Eagle and Martinstown, as well as in many parts of the Marshwood Vale. In Somerset John Billingsley wrote in 1708 of the watered meadows in the parishes of Crowcombe, Stogumber, Monksilver and Nettlecombe in the Quantock and Brendon Hills and noted that they 'are as good as any in the county... and invaluable for keeping stock throughout the year'. Further west in Somerset there are remains at Tolland, Dunster and Dulverton, while a complete hillside system survives at East Nurcott Farm, Winsford on the border of Exmoor where the river Quarme is brought along the steep hillside by a leat and made to overflow through a series of subsidiary gutters. Similar hillside meadows are to be found in many parts of Devon, notably in the South Hams and around Torrington. Charles Vancouver, writing in 1808, commented on the number of such meadows throughout Devon and stated that 'the practice obtains very generally through most of the vallies in the county' (Vancouver, 1808, pp. 313-22). Billingsley described the early grass produced by the catchwork meadows in Somerset and wrote 'a great part of the watered lands lie on steep declivities; and as the water passes quickly over them, and never lies stagnant, not a rush can be seen' (Billingsley, 1798, pp. 264-5).

The most remarkable example of the use of catchwork meadows occurred on Exmoor. This vast royal forest comprising some 22,400 acres (9,000ha) of bleak, uninhabited moorland was purchased from the Crown in 1815 by the wealthy Midlands ironmaster, John Knight who paid £50,000 for it. In spite of Knight's enthusiasm and expenditure, his large-scale attempts to reclaim the moorland did not achieve the success he had anticipated. In 1841, however, his son, Frederick Winn Knight, took over the management of the property and controlled it until his death in 1897. He was much more successful, largely through building several farmsteads on the moor, letting them at low rents and encouraging the tenants to undertake reclamation of the moorland. A major weapon in the struggle to create good agricultural land and nutritious grazing out of the hilly terrain and the acid, peaty soil was the extensive use of catchwork meadows. These had already been introduced on various parts of the moor by John Knight, and their use was continued and extended by his son and especially through the influence of the steward, Robert Smith. Smith served as steward from 1848 to 1861 and remained as a tenant on Exmoor until 1868; he was an enthusiast for catchwork water meadows, and became an expert in their construction and management, conducting tests on the temperature and quality of the springs and on the best methods of using the water. In 1851 he wrote an account of his work on the meadows which was published in the *Journal of the Royal Agricultural Society* (Smith, 1851). Smith described the meadows he had created, the main 'carriage' or water course, the gutters along the hillside and the calendar of watering. As with the floated meadows, he claimed that 'the first outlay is the main expense ... the annual expenses to maintain their efficiency being small in comparison with the result'. In suitable locations he arranged for the main leat to run through the farmyard or past cattle sheds so that dung and urine were mixed with the water and provided additional nutrients. The meadows were laid out in 5- or 6-acre (2- or 2.5-ha) sections, and proved invaluable in producing early grass for the ewes and lambs and abundance of hay for the winter. Smith claimed that the water meadows provided by the landlord were greatly appreciated by the tenants and that the system was spreading rapidly on the Knight's Exmoor estate, 'new meadows are being laid out upon nearly every farm'. By the 1880s rents per acre on Exmoor ranged from 5s to £1 5s, while water meadow land could command as much as £3 10s per acre. The remains of these meadows are still very evident on the Exmoor hillsides (Orwin, 1929, pp. 31-79; Havinden, 1981, pp. 108-9).

—THE DECLINE AND ABANDONMENT— OF THE MEADOWS

During the later nineteenth century the water meadows began to fall into disuse, and by the mid twentieth century only a few were still being worked. The introduction of artificial fertilisers meant that the fold of the sheep flock was no longer essential for growing satisfactory crops of corn on the downlands, and the number of sheep declined rapidly. For example, sheep numbers in Dorset declined from more than 700,000 in 1870 to 300,000 in 1900, and to 46,000 in 1947. In Wiltshire sheep numbers fell from 775,000 in 1870 to 162,000 by 1939. The widespread use of imported feedstuffs, together with root and fodder crops and the introduction of new strains of grass, meant that the early grass provided by the water meadows was no longer so important. Moreover, the water meadows were labour intensive: hatches, channels and drains needed constant attention, and the surface could not bear the weight of modern machinery nor of twentieth-century tractors, so hay had to be cut using scythes. Above all, the prolonged agricultural depression starting in the 1870s brought profound changes to farming. The demand for early lambs declined in the face of imports from New Zealand; the price of wheat and barley fell dramatically because of foreign imports; and more and more land was laid down to grass. It was only the sale of liquid milk, which could now be sent to the towns by rail, which kept many farmers in business. The abandonment of the traditional sheep/corn husbandry in favour of dairy farming led to a further decline in the acreage of water meadows.

All the chalkland valleys, and many hillsides in other parts of the West Country, have the easily recognisable hatches and intricate network of channels, but most present a sorry picture of neglect and decay. Ditches are filled and channels flattened for the convenience of modern farming operations. In only a few places have water meadows survived and continue to be worked in the traditional manner. Examples can be seen in Dorset at Wolfeton on the Frome north of Dorchester and on the Dewlish Brook east of Puddletown. In Wiltshire meadows are still floated along the Avon at Lower Woodford and, as we have seen in the last chapter, at Britford south of Salisbury. Sluices, hatches and channels survive in the Wey valley between Bramshott and Headley, in the Meon valley at Meonstoke, and along the Itchen, Test and Avon valleys. A selection of the tools used by the watermen or 'drowners' can be seen in the Countryside Museum at Breamore, Hampshire. Remains of the now disused systems can also be seen along most of the chalk streams of Wiltshire and Dorset, notably at West Overton and Clatford on the Kennet, at Bishopstone on the Ebble, and at Fordington, Moreton, Puddletown and Piddlehinton along the Frome and Piddle in Dorset.

The value of the meadows as a cost-efficient method of producing grass without the use of expensive modern fertilisers is still recognised, and attempts have been made to revive their use or adapt them to modern conditions. During the 1980s an experiment of this kind was made by the Ministry of Agriculture on the river Ebble at Odstock near Salisbury. The hatches of a disused water meadow were restored, but instead of the surface of the meadow being ridged to accommodate channels and drains it was flattened but carefully 'graded', or sloped, so that the water entered at one side and gently flowed across the surface of the meadow to the drain at the other side. It was hoped that this would allow machinery to be used without damaging the surface. Sadly, the scheme did not prove satisfactory and the experiment has been abandoned.

A much larger and more elaborate scheme has been introduced by the National Trust on the Sherborne estate in north-east Gloucestershire. More than 150 acres (60ha) of water meadow were constructed along the bank of the river Windrush at Sherborne during the Napoleonic War period or soon after. The meadows continued in use until the early twentieth century, but were ploughed up, the ridges destroyed and the surface of the meadow flattened to grow corn during the Second World War. During the 1990s the Sherborne water meadows were restored by the National Trust. The stone sluices have been rebuilt, strongly made wooden hatches installed and the deep channels which convey the water from the Windrush have been excavated. No attempt has been made to restore the ridges which formerly spread the water over the land, but the flat, gently sloping surface is expected to distribute the water evenly and ensure that it is kept moving. Already the meadows are producing lush grass and abundant hay, and have also become a valuable haven for wildlife, especially for waterfowl.

CHAPTER 13

Inappropriate technology?
The history of 'floating' in the North and East of England

SUSANNA WADE MARTINS AND TOM WILLIAMSON

— INTRODUCTION: A SLOW ADOPTION —

As we saw in the previous chapter, irrigated water meadows had become a normal part of husbandry throughout the southern chalklands of England, and across much of the West Country, by the early years of the eighteenth century. Floating was well established even in areas in which catchwork systems had to be employed, and in which streams rising in non-calcareous geologies had to be utilised – circumstances in which the practice was probably less effective. Yet a century later commentators were unanimous in their belief that floating was virtually unknown in the northern and eastern regions of England. According to the prominent agricultural writer Arthur Young, in Norfolk in 1804 the practice was 'of very late standing' and 'the experiments made are few' (Young, 1804a, p. 395). In Suffolk in 1813 there was 'not a well-watered meadow' (Young, 1813a, p. 196); while in Lincolnshire, Buckinghamshire and Cambridgeshire the authors of the various *General Views* of 1813 noted only single examples in operation (Priest, 1813, p. 281; Young, 1813b, p. 312; Gooch, 1813, pp. 258-62). In Kent, according to Boys in 1813, the practice had 'as yet very few friends' (Boys, 1813, p. 164) and in the same year Middleton asserted that 'irrigation makes no part of the practice of a Middlesex farmer' (Middleton, 1813, pp. 398-9). Similarly in Huntingdonshire in 1813 it was said that 'artificial watering of meadows or other low lands is in a very backward state' (Parkinson, 1813, p. 242).

— FLOATING IN THE SEVENTEENTH AND — EIGHTEENTH CENTURIES

There are, however, indications that irrigation had been adopted in some of these eastern areas during the course of the seventeenth and eighteenth centuries, albeit on a limited scale. Irrigated meadows were thus established at Babraham in south Cambridgeshire between 1652 and 1654, by the lord of the manor, Sir Thomas Bennet (Rosen, 1978); and several early nineteenth-century accounts describe how, during the creation of irrigation systems, older works were uncovered. Thus when Thomas Purdey began to irrigate the meadows at Houghton St Giles in Norfolk in 1804 he found, according to Young,

> the foundations of an old sluice ... in a sound state; and the whole immediately renewed; on further examination, the carriers and drains in the meadow were all traced, opened afresh, and thus an irrigation formed on very nearly the plan of old works, which had been utterly neglected for at least eighty years (Young, 1804a, p. 399).

Sporadic attempts were, therefore, made to introduce floating into eastern England in the course of the seventeenth and eighteenth centuries, but these were seldom very successful, or very long-lived.

Some of meadows that were created in these areas, moreover, seem to have been managed in rather different ways to the familiar Wessex examples described in the previous chapters. Two noted nineteenth-century advocates of floating, William Smith and Arthur Young, were thus perplexed by the examples at Babraham in Cambridgeshire. Smith noted that 'the form of the works seems to prove that they were not designed by any person from Wiltshire, and that the possessors are totally unacquainted with the management and utility of water meadows' (Smith, 1806, pp. 116-17). In Young's words: 'there does not seem to be the least intelligence or knowledge of the husbandry of water. No other art is exerted, but that merely of opening in the banks of the river small cuts for letting the water flow on to the meadows' (Young, 1797, p. 177). More surprising still, irrigation began here on Easter Monday 'and never sooner than two weeks before': in other words, the meadows were designed to produce a large hay crop, and not – as was principally the case with the Wessex meadows – an 'early bite'.

All this amounted, in Young's opinion, to 'circumstances ... not easy to understand'. Yet other meadows in the East were managed in a similar way. Thus the extensive irrigated meadows between

Cassiobridge and Watford in Hertfordshire, and around Uxbridge in Middlesex, were mown twice a year, but only the aftergrass was fed (Young, 1804b, pp. 178-9). Elsewhere descriptions of meadows in the seventeenth and eighteenth century are often sketchy or ambiguous, but it is possible that many others were managed primarily to enhance the summer hay crop, rather than to provide early spring feed – a situation clearly at variance with Wessex practice, where the 'early bite' was the pre-eminent motive for irrigation.

— Floating in the early nineteenth — century

The later eighteenth and early nineteenth centuries saw a remarkable upsurge of interest in the practice of floating in the North and East of England. This was not, however, the result of the direct diffusion of the practice from its old heartlands, gradually spreading as neighbour copied the example of neighbour. On the contrary: wherever water meadows were adopted it was almost always either by, or with the active encouragement of, major aristocratic 'improvers', and the systems themselves were designed by experts specially brought in from the South and West of England. Floating may thus be viewed as an alien imposition rather than an organic development in farming practice. Thus we find that in Bedfordshire the technique was introduced shortly after 1800 by the Duke of Bedford, but by 1813 it was still limited to a few places near the Ducal residence at Woburn Abbey (Batchelor, 1813, pp. 484-93). In Leicestershire, similarly, floating was restricted to the north-east of the county, and exclusively associated with the activities of Robert Bakewell and his clique of improvers (Pitt, 1813, pp. 201-2).

In this context it is noteworthy that in 1802 the Board of Agriculture began offering an irrigation premium 'to the person who shall in a county where irrigation is not generally in practice, water the greatest number of acres and in the completest manner': one of the earliest examples of government grant-in-aid for an agricultural improvement. To claim the premium the applicant had to provide 'an account of the old and new state of the land, and of its value, and of the method, expense, and produce verified by certificate, to be laid before the board by the third Tuesday in January' (*Annals of Agriculture* 37, p. 68). Both gold and silver medals were offered, for many years. The Board's example was soon followed by private landowners. The Duke of Bedford, and Thomas William Coke of Norfolk, both presented prizes for irrigation projects at their 'sheep shearings' (agricultural shows), held at Woburn and Holkham respectively. In 1803 Coke, under the aegis of the Norfolk Agricultural Society, was offering 'a piece of plate to the value of five guineas to such persons as shall convert the greatest area of waste or unimproved meadows into water meadows in the completest manner'(*Annals of Agriculture* 39, p. 322). Tenants might be encouraged in more direct and orthodox ways. James Loch described in 1820 how the Marquis of Stafford's tenants were allowed the rough materials to construct the floodgates, and 'the example was shown them of what could be done in this respect at Trentham... where a new water meadow at the Home Farm costing £20 per acre' had been constructed (Loch, 1820, pp. 199-200).

Educated interest in the late eighteenth and early nineteenth centuries was mirrored in the appearance of a number of publications, including in particular William Smith's *Observations on the Utility, Form and Management of Water Meadows*, published in Norwich in 1806 (Figure 13.1). Smith was a civil engineer and pioneer geologist from Gloucestershire, who was later to earn the epithet of 'father of English geology'. He undertook irrigation and drainage schemes for both the Duke of Bedford, and for a number of Thomas Coke's improving tenants, laying out particularly extensive and complex systems at Wighton and West Lexham in Norfolk. The *Observations*, dedicated to Thomas William Coke, was clearly intended to attract further custom in the area, at a time when that noted improver was actively encouraging irrigation.

Norfolk is one area in which the spread of floating in the nineteenth century has been studied in some detail (Wade Martins and Williamson, 1994). Only three irrigated meadows are known to have been established in the county before 1800: at Riddlesworth (1792), Thetford (c.1793), and at West Tofts. All were located in the sandy Breckland district, and all were the work of prominent 'improving' landowners (Wade Martins and Williamson, 1994, pp. 21-3). The following decade saw the establishment of a number of new systems, mainly in the valleys of the rivers Nar and Stiffkey in west Norfolk: at Wighton (1802), Lynford (1803), Billingford (1804), Houghton St Giles (1804), Raynham (c.1806), Taverham (c.1806), Beachamwell (1806), West Lexham (1806), Waterden (1808), Castle Acre (1810) and Heacham (1810) (Wade Martins and Williamson, 1994, pp. 23-7). A number of other undated examples in the county, like those at East Lexham and Easton, were probably constructed at around the same time: only one securely-dated example - a small catchwork at Mileham,

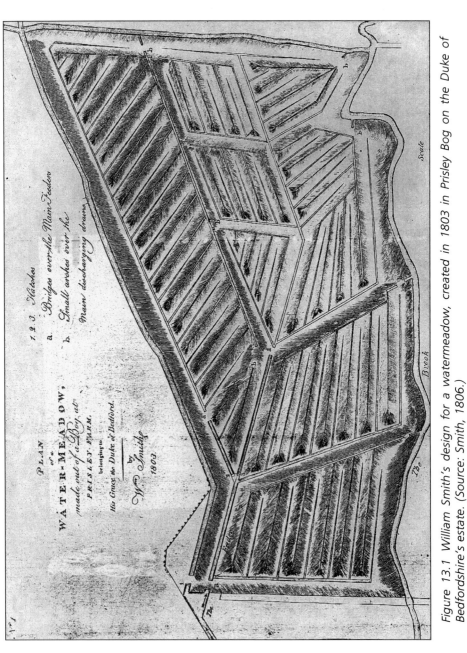

Figure 13.1 William Smith's design for a watermeadow, created in 1803 in Prisley Bog on the Duke of Bedfordshire's estate. (Source: Smith, 1806.)

created in 1816 – was certainly established after 1810. Interest in floating thus coincided with the Napoleonic Wars, a period in which agriculture was prospering, farm incomes growing, and optimism was in the air. Almost all known irrigation schemes in the county were, moreover, created by tenants of large Holkham estate farms, or by members of Thomas William Coke's clique of enthusiastic improvers.

The Norfolk water meadows were, for the most part, complex and elaborate – even flamboyant – creations. One of the best preserved is that at Castle Acre, in the valley of the river Nar in West Norfolk (Figure 13.2). It was constructed in or soon after 1808 by Thomas Purdey, a leading Holkham tenant: in 1810 he was awarded a silver teapot, basin, and cream ewer at the Holkham sheep shearing for irrigating 30 acres (12ha) here (*Norfolk Chronicle*, 30 June 1810). His meadows, like most created in Norfolk at this time, broadly resembled Wessex examples, in that they consisted of a number of blocks of 'beds' or 'panes', c.8m wide and up to 80m in length. But the sluices were substantial brick constructions, and two aqueducts took the principal carriages across the river. The whole system probably cost in excess of £800. The meadows were principally fed from a leat, taken off the river Nar nearly 0.5km upstream, but the system also included a section of catchwork, fed from springs and by the run-off from field ditches (Wade Martins and Williamson, 1994, pp. 25, 27–31).

The Castle Acre meadows were flourishing in 1844, when they were described in glowing terms by Richard Bacon in his *Report on the Agriculture of Norfolk* (Bacon, 1844, p. 290). The same writer also refers to the systems at West Lexham, Wighton and Kempstone, but not to any of the others created in the county during the Napoleonic War period, and such evidence as there is suggests that most had been abandoned by this time. The Castle Acre meadows seem to have remained in operation into the 1880s, but there is no evidence that those at Lexham, Wighton or Kempstone continued in use as late as this.

This flurry of early nineteenth-century activity evident in Norfolk was shared by several other areas in northern and eastern England, to judge from the *General Views* and other sources. By 1813 it was reported that even in Westmoreland irrigation was 'practised on a small scale – always with great success', while in Lancashire a Mr Gawthorpe had irrigated 5 acres (2ha) with water 'from the top of a mountain' and increased his hay crop five fold (Bailey and Culley, 1813, p. 325; Stevenson, 1815, p. 510). The 'boom years' of the Napoleonic War period were, however, followed by slump, and only

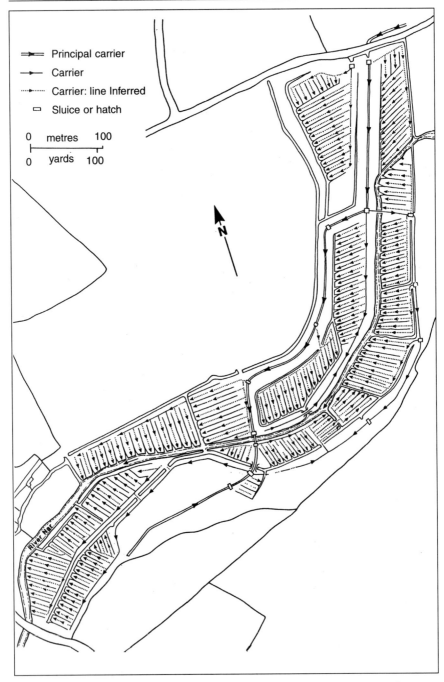

Figure 13.2 Reconstructed plan of the Castle Acre water meadows, Norfolk, based on ground survey and on the RAF 1946 aerial photographs.

as agriculture began to recover in the 1830s, in the period of capital-intensive 'high farming', did interest in water meadows begin to revive in the North and East.

— THE 'HIGH FARMING' PERIOD —

The Royal Agricultural Society of England was founded in 1838 and the first volume of its *Journal* appeared in 1839. Its pages contained much discussion of the benefits of floating, and of the advantages and disadvantages of various methods, throughout the 1840s and 1850s. Philip Pusey, the editor of the *Journal*, was a particularly enthusiastic advocate. He described irrigated meadows in glowing terms, as:

> The triumph of agricultural art: changing as it does, the very seasons ... a slight film of water trickling over the surface rouses the sleeping grass, tinges it with living green amidst snows or frosts, and brings forth a luxuriant crop in early spring, just when it is most wanted while other meadows are bare and brown. (Pusey, 1849, pp. 462–79).

Pusey himself succeeded to his Berkshire (now Oxfordshire) estate in 1828 and soon installed a system of meadows which enabled him to quadruple his flock of sheep. One 20-acre (8-ha) meadow, he claimed, kept 400 sheep in feed for five months, and the profits from the land were thereby increased by 30 per cent (Caird, 1852, p. 112). Several other systems are described in the pages of the *Journal*, all well away from the chalklands of the South. The extensive and complex system of irrigation installed on the Duke of Portland's estate between Mansfield and Ollerton, in Sherwood Forest in the 1820s were warmly praised by John Denison in 1840 (Denison, 1840). Three hundred acres (120ha) were watered by means of at least 7 miles (11km) of flood dyke: the total costs of the improvement, which included some drainage, amounted to no less than £38,000.

Many of the schemes described in the pages of the *Journal* were in the uplands of Britain. A meadow in the Brendon Hills is described, and another on a Scottish sheep farm (Roal, 1846). The complex systems on Exmoor, created by the Knights in the 1840s as part of their ambitious reclamation scheme, have been described by Bettey in the previous chapter. Water meadows, according to the ever-optimistic Pusey, were the 'talisman by which a mantle of luxuriant verder might be spread across the mountains and moors of Wales and Scotland, of Kerry and Connemara' (Pusey, 1846).

Throughout the nineteenth century agricultural encyclopaedias continued to enthuse about the practice, and while they acknowledged that it was most widely employed in the South and West they pressed its suitability for other areas. Nevertheless, in spite of all this improving rhetoric, Copland's *Agriculture Ancient and Modern* of 1866 pointed out that while in the southern chalklands and the West Country irrigation had been used 'time out of mind' the practice had still not spread significantly beyond these regions (Copland, 1866, p. 762). Archaeological evidence bears this out. Thus in the East of England the distinctive earthworks of 'bedworks' are remarkably rare: only a handful of examples appear on the RAF 1946 aerial photographs for Suffolk, Norfolk, or Essex. Copland, moreover, was writing at the time of the maximum spread of water meadows. By this time, various agricultural innovations were already rendering them, if not unnecessary, then certainly less attractive as a form of investment. The onset of agricultural depression in the late 1870s finally ended these various experiments at floating in the North and East.

— Explaining divergence —

The history of water meadows in the East and North of England is thus very different from that in the South and West. Irrigation never became a central feature of husbandry here, even in those districts (such as western East Anglia or the Lincolnshire Wolds) in which farming systems of sheep-corn type, little different from those of the Wessex chalklands, were followed. Irrigation was only sporadically adopted in the course of the seventeenth and eighteenth centuries, and was sometimes used exclusively to improve the summer hay crop, rather than to force an early growth of grass. Interest blossomed in the early years of the nineteenth century, when agriculture was flourishing, farm incomes and rents high, and farmers and landowners in a particularly optimistic mood. A number of systems, clearly managed along conventional Wessex lines, were adopted at this time, largely under the influence of aristocratic improvers. Nevertheless, uptake remained slow, and in the course of the century many of the meadows constructed with such enthusiasm in the Napoleonic War years were abandoned, although right through the 'high farming' period interest in floating remained strong among agriculturalists, and a number of new systems were created.

Several reasons for this divergent pattern of development can be suggested. The first is that in many of these regions the topography was simply unsuitable for floating. In much of the East of England, especially in Norfolk, Suffolk, and Essex, the gentle gradients and wide floors of many valleys ensured that it was impossible to implement cheap catchwork systems, and that bedworks could only work effectively if they were supplied by a long leat: only then could a sufficient head of water be obtained, as Kent pointed out in 1794 (Kent, 1794, p. 11). This could greatly increase the costs of construction: Bacon thus asserted in 1844 that floating had only been introduced into Norfolk 'at so great an expense as to prevent it being carried out to any considerable extent' (Bacon, 1844, p. 291). Equally important was the fact that extended leats led to practical problems of implementation, for the consent of other proprietors, owning lands 'between those to be flooded and the place where the water must be taken out of the river', had to be obtained (Smith, 1806, p. 120). The greater the length of the leat, moreover, the greater the likelihood that it would interfere with the water supply to mills – a source of perennial disputes, as readers of *The Mill on the Floss* will recall. Young and others repeatedly bemoaned the way that milling operations interfered with irrigation schemes.

Topography might have discouraged irrigation in other ways. The flat, wide valleys found in many parts of the East, and especially in East Anglia, can locally contain areas of poorly draining, peaty soils, even if the river itself arises in calcareous springs. As Smith and others explained, the installation of irrigation schemes – somewhat paradoxically – improved the drainage and thus the quality of the grass on such land; but nevertheless, descriptions of irrigated meadows found in the *General Views* and elsewhere sometimes complain about the coarse nature of the herbage in irrigated meadows.

But while topography was undoubtedly one factor that discouraged the spread of the system, it was not the only one: many areas in the East of England, in the Lincolnshire Wolds for example, have valleys with profiles and soils little different from those found in the Wessex chalklands. Of equal importance was the climate. As Pusey observed in 1849, irrigation was less well suited to the drier and colder areas of the country. Lack of water was seldom a problem, even in the eastern counties, during the winter and early spring, but low spring temperatures could be critical. In Pusey's words, 'a slight difference in warmth has a marked effect in hastening or retarding growth' (Pusey, 1849, p. 477). Even an ardent advocate like William Smith was forced to admit that, at Lexham in Norfolk, irrigation in the second year failed to provide much in the way of an 'early bite' because of the frost: 'the water was applied every night, but from the coldness of the weather produced but little effect' (Smith, 1806, p. 115).

Warming the ground may not have been the only benefit bestowed by irrigation, but it was probably the main one. As Cutting and Cummings show (see Chapter 11), increases in temperature above 5°C are crucial in effecting the performance of grass growth. In the more continental East of the country, and in the North, sharp frosts in March and February tended, in many years, to override any warming effects supplied by the water; and in the worst cases, the water could actually freeze, causing considerable damage to the grass. While it was certainly possible to force an 'early bite' by irrigation in many of these areas, it was difficult to do so on a regular basis. As an illustration, during the period 1941–70 the average number of winter degree-days below 0°C (i.e. the accumulated daily temperatures below freezing, November to February) were 160 in Norfolk, 130 in Kent, and 125 in Surrey; the area around Salisbury, in contrast, enjoyed only 95 (Smith, 1976). All this explains, of course, why some of the irrigation systems established at a relatively early date in the East (like those at Babraham) were simply intended to enhance the summer hay crop, rather than force an early 'bite'.

The practice of irrigation in such cases was thus similar to that in neighbouring parts of the European mainland, especially Scandinavia (Emanuelsson and Moller, 1990, p. 136). Here irrigation of meadows was quite commonly carried out by the nineteenth century, but seldom to force an early growth of grass. However, in an English context the benefits to be gained from increasing the hay crop were in most cases insufficient to justify the expense of constructing irrigation systems, given that – for reasons of topography – their costs were often greater here than in the South and West. By the early years of the eighteenth century, moreover, rotations featuring regular courses of fodder crops – principally turnips and clover – were in widespread use in eastern arable areas, and the need for meadow hay correspondingly less.

– Discussion –

It might appear that the history of floating outside its traditional heartlands is a subject of only antiquarian interest. In fact, it has a number of important lessons for students of agriculture and technical innovation. Firstly, it demonstrates clearly the ways in which quite minor variations in the character of the natural environment could have a profound impact, both directly and indirectly, upon the development of farming systems. Not only did barely noticeable idiosyncrasies of climate and topography discourage the adoption of floating in the arable East of England; in addition, this in turn probably encouraged the early use of turnips and clover in these regions, thus putting their arable farming systems on a trajectory which continued to distinguish them from the South of England until well into the nineteenth century (Bowie, 1990). Secondly, and of more importance, is what the history of floating tells us about the nature of innovation, and of agricultural discourse, in the periods of the 'agricultural revolution' and 'high farming'. Much of the agricultural history of the eighteenth and nineteenth centuries is still written on the basis of the texts produced by Arthur Young and other contemporary 'experts'. Yet it is not always clear how much these individuals really knew about the practicalities of agriculture in the various districts they described. They bemoaned the failure of farmers to adopt irrigation, ascribing this to a mixture of laziness and ignorance: Copland for example asserted that the failure of irrigation to spread was simply due to a 'want of knowledge' on the part of the farming community (Copland, 1866, p. 762). What was really needed, most commentators agreed, was the example and encouragement of enlightened improvers. Thus Parkinson in the *General View* for Huntingdonshire, lamenting the absence of floating, commented:

> What a pity that some nobleman or gentleman would not cause this improvement to be adopted in a few parishes, and then the example would cause farmers and others to adopt the same plan in similar situations, for their own advantage (Parkinson, 1813, p. 313).

But where aristocratic improvers led, the farming community seldom followed: and for good reason. Floating was simply not economically viable in most contexts outside Wessex and the West.

The story of floating actually forms part of a wider pattern: for it can be argued that many aspects of the 'agricultural revolution' of the late eighteenth and nineteenth centuries should be considered in social and ideological as well as in agrarian and economic terms. Improvement, and especially land reclamation, was an aristocratic duty: a proclamation that nature could be transformed by the application of knowledge, and thus an affirmation of the rights of the social elite of large landowners and capitalist farmers. Improvement gave legitimacy to land ownership and the farming interest, at a time when the accumulation of property in the hands of the few, and restrictions on the import of foreign grain, were both being questioned by an increasingly assertive urban middle class. As Pusey himself put it, describing the potential benefits of irrigation in upland areas:

> If the plain means of improvement and employment are still neglected, it will be possible to tax owners of needless deserts with supiness; and difficult to deny that they hold in their hands more of their country's surface than they are able to manage for their own good or for that of the community (Pusey, 1846).

Contemporaries thus singled out for particular comment irrigation schemes carried out as part of wider projects of reclamation, such as the 300 acres (120ha) of meadows on the Duke of Portland's estate between Mansfield and Ollerton, in Sherwood Forest. Here, according to Denison, 'the contrast between the wild beauties of nature and the finished works of cultivation and art, thus placed side by side, is very striking and remarkable'; while according to Burroughes the system demonstrated 'art, science and the ingenuity of man crowned with triumphal success' (Denison, 1840, p. 360; Burroughes, 1846, p. 60). The use here of terms familiar from the writings of Repton, Loudon and other landscape gardeners is noteworthy: and indeed, throughout this period the term 'improvement' was used indiscriminately for both park-making and land reclamation schemes. In this context, it is worth emphasising how many examples of floated meadows in the North and East were associated with designed, ornamental landscapes. Thus at West Lexham in Norfolk the diminutive lake created in front of the large farmhouse was clearly intended in part to provide it with a park-like setting, appropriate to a capitalist farmer of several hundred acres, but it also provided the main supply of water to the irrigated meadows here. In the adjacent parish of East Lexham, the meadows lay in the centre of the park, in full view of Lexham Hall, and drained into the ornamental lake known as the Long Water. At Claughton Hall in Lancashire, 10 acres (4ha) were irrigated from a canal in the park; at Woburn in Bedfordshire the meadows within the park were fed from the Temple Reservoir. Even when they were not actually located within landscape parks, irrigated meadows could be considered objects of beauty – embellishments to the landscape – as much as practical pieces of agricultural technology. According to James Loch, the Marquis of Stafford's meadow at Trentham was 'adorned by fine trees ... it forms a very perfect union of useful and ornamental farming' (Loch, 1820, p. 200).

The history of 'floating'

The story of irrigation outside its western and southern heartlands is thus of some interest to historians and others. The principal lesson of that story – that the adoption of an innovation does not necessarily reflect a careful and balanced judgement of agricultural and economic realities on the part of the social elite, but can have much more varied and complex motivations – is doubtless relevant in many other contexts.

CHAPTER 14

Historical changes in the nature conservation interest of the Fens of Cambridgeshire

CHRISTOPHER NEWBOLD

The history of fen drainage in East Anglia has been described in some detail in earlier chapters. This section deals with the environmental effects of this momentous change in the landscape, a subject generally glossed over in conventional archaeological and historical accounts.

In AD 970 Abbot Floriacensis from the abbey on the Isle of Ely described East Anglia as being encompassed to the south and east by the ocean, and to the north by the moisture of large fens which, 'arising in the heart of the island descend in great rivers to the sea, the inner parts being a rich soil and famous for pasturage which presented in summer a most delightful green prospect' (de Gray Birch, 1881). The monk Felix from the abbey at Crowland, 80 km north-west of Ely, wrote in the eleventh century of a landscape of reeds as far as the eye could see, with hidden lakes, wooded islands and a waste of mere and quaking bog. In the following centuries the area of drained land was extended, principally on the silt soils on the seaward side of the area, towards the Wash (see Chapter 9) but throughout the Middle Ages the peatlands in the interior remained largely undrained, and a rich and varied range of natural environments survived.

The distinctive economy of the Fens was a significant factor in moulding the character of these varied environments. In summer those parts not flooded were used as hay pasture, with the aftermath grazed by sheep or cattle; the wetter areas were simply grazed by cattle. With peat for fuel, reed to thatch the cottages, osiers for baskets and traps and eels, fish and wildfowl for food this was a truly sustainable self-supporting economy, with a diversity of habitats such as fen meadow; wetter fen, for grazing, reed, and sedge; osier beds; bog, lake and river. The local population unwittingly sustained and managed the Fens in what would be regarded today as a large nature reserve.

A verse printed in the book *Polyolbien* in 1662 describes their life:

> The Toyling fisher here Tewing [hauling] of his net
> The fowler is imployed his lymed twigs to set
> One underneath his horse to get a shoote doth stalk
> Another over dykes upon his stilts doth walk
> There others with their spades the peats are squaring out
> And others from their Carres [wet woodland] are busily about
> To draw out sedge and reed for thatch and stoves [ridge] fit.

Limited areas of the peat fens had been drained in medieval times, but only in the seventeenth century were large-scale schemes of improvement instituted. Arguably the first 'government'-funded drainage scheme began in 1630 when King Charles I asked the fourth Earl of Bedford to drain the Fens. The Earl owned 8,000ha at Thorney, 16km east of Peterborough. A Dutchman, Cornelius Vermuyden, was appointed Chief Engineer (Dugdale, 1662). Several new drains were cut, of which the most well known is the Old Bedford river, which stretches for some 34km from Earith in the south to Salters Lode in the north. This drain, with sluices at either end, took water direct to the sea, thus relieving the meandering course of the Ouse by an intervening washland which flooded in winter receiving water from the rivers Lark, Little Ouse and Wissey.

The principal object of the scheme was to make the land more fit for summer grazing, although some arable land use was also envisaged. But livelihoods were threatened; some landowners felt the fertility of the soil would be affected, as winter floods 'warped' and enriched the land with fine silts. A contemporary account, the *Powte's Complaynte*, sums up the feeling at that time, and begins:

> Come brethren of the water and let us all assemble
> To treat upon this matter which makes us quake and tremble
> For we shall rue it, if't be true the Fens be undertaken
> And where we feed in Fen and reed they'll feed both beef and bacon.

Despite riots, the scheme went ahead. The Fens began to change, although there were several reverses due to the subsidence of the surface, as the peat became so dry that it shrank, powdered and blew away. Drainage windmills (the 'silent spectators of the floods' as they

Figure 14.1 A Panorama of drainage windmills, North Drove, Deeping in 1828.

were sarcastically called) were only just able to keep pace with deteriorating conditions (Figure 14.1): it was advent of the steam pump that changed the Fens for all time. Recommended for use in 1789, the steam engine was first used in the Cambridgeshire Fens in 1830. The Fens now became an increasingly arable district, and although sheep and cattle were still grazed on pastures, especially along the banks of the main arterial drains (Miller and Skertchley, 1878), ploughland replaced pasture and reed grounds on a grand scale.

Ironically, the washland created by Vermuyden to hold flood waters between the Old Bedford river and the New Bedford river is now an internationally important reserve for wildfowl, known as the Ouse Washes. The ditches hold a relict flora and fauna once common in the Fen ditches and on the edges of the meres and lakes. Relict fens, such as Holme Fen National Nature Reserve (NNR), Woodwalton Fen NNR, Wicken Fen, a National Trust Nature Reserve, and Chippenham Fen NNR give us glimpses into past floras, when such areas were more extensive (Ratcliffe, 1977; Newbold et al., 1989).

— THE WILDLIFE OF THE FENS —

Descriptions of the Fens by monks in the eleventh and twelfth centuries, and by later historians, coupled with what exists today as true fen remnants, now found in nature reserves, enables historical botanists to reconstruct the plant life found in fen meadow, fen, raised bog, river and mere.

The most common habitat on the drained ground would have been damp meadow; less common would be flood meadow and what is known today as fen meadow. In damper hollows on spring lines around the edge of the Fens, true fen habitat would be found. In undrained areas, often below sea level, reed bed and mere would be used for wildfowling and fishing. The meres survived into post-medieval times. Ramsey, 24km south-east of Peterborough, was famous for its wildfowl market which 'exported' produce to London. Gradually, as drainage scheme followed drainage scheme, all these meres became agricultural land. The last great mere, Whittlesey Mere, some 8km long and 0.8–6.5km wide was finally drained in 1851. The raised bog by Whittlesea Mere, now Holme Fen NNR, was starved of water and began to shrink. The well-known fen posts recorded the loss of the bog with the peat shrinking some 4m over the century (Godwin, 1978).

Apart from the meres (which remained more or less intact until the early seventeenth century) the areas of damp meadow and fen meadow would both have fluctuated in area, dependent on the efficiency of drainage in combating the vagaries of tide and weather. Most fen meadows would anyway have been flooded in winter before the mid-seventeenth century. In some areas – particularly on former gravel islands – woods of oak and silver birch developed, fringed on the wetter edges by alder carr. So we have a picture of a very varied fen landscape, of reed bed, river and mere raised bog, true fen, fen meadow, flood meadow, damp meadow, wet woodland or carr and dry woodland.

True flood meadow is now a rare habitat in England, of which less than 500ha survives. The only relict community near to the Fens surviving today is that at Portholm close to Huntingdon. The main plants are the great burnet, a blackberry-like flower of pinkish burgundy colour, which here grows alongside the grass meadow foxtail. Nodding bell-like heads of the purple snake's head fritillary, interspersed with the yellows of buttercups, would abound in the spring. There is every reason to suppose that these communities would have been reasonably commonplace along the flood meadows of all the rivers which meandered through the Fens.

Plant communities are defined according to the National Vegetation Classification (NVC): the principal fen communities are described in Table 14.1. It should not be imagined that all flood plains would have supported such rich pasture and such an abundance of colour. Some soils would have been so wet that soft rush and the Yorkshire-fog grass would have produced poor damp meadow (see Table 14.1, the MG10 community).

TABLE 14.1 Requirements of typical wet grassland plant communities (MG4–5, MG10–13)

National Vegetation Classification	Dominant species	Occurrence (habitat)	Soil type	Watertable
Flood meadows: meadow foxtail and great burnet grassland (MG4).	Meadow foxtail (*Alopercurus pratensis*), common mouse-ear (*Cerastium fontanum*), crested dog's-tail (*Cynosurus cristatus*), red fescue (*Festuca rubra*), meadowsweet (*Filipendula ulmaria*), Yorkshire-fog (*Holcus lanatus*), great burnet (*Sanguisorba officinalis*) autumn hawkbit (*Leontodon autumnalis*), perennial rye-grass (*Lolium perenne*), meadow buttercup (*Ranunculus acris*), common sorrel (*Rumex acetosa*), dandelion (*Taraxacum officinale*), red clover (*Trifolium pratense*) and white clover (*Trifolium repens*).	River floodplains in central and southern England. In the past, reasonably common on the river floodplains of the Fens. Now rare, found in less than 500ha.	Alluvial (neutral or calcareous).	Soils moist to very locally damp. Free-draining above, or sometimes waterlogged (and gleyed) at depth.
Old grazed hay meadows: crested dog's tail and common knapweed grassland (MG5).	Common bent (*Agrostis capillaris*), sweet vernal grass (*Anthoxanthum odoratum*), common knapweed (*Centaurea nigra*), crested dog's-tail, cock's-foot (*Dactylis glomerata*), red fescue, Yorkshire-fog, common bird's-foot-trefoil (*Lotus corniculatus*), ribwort plantain (*Plantago lanceolata*), red clover and white clover.	Widespread in lowlands of Britain – especially in the clay farmland of English Midlands. Often on ridges of ridge-and-furrow. In the past a common community found at the edge of the Fen basin.	Circumneutral brown soils but over a wide range (with consequent variation in the make-up of flora).	Soils moist. Where soil particles are finer, drainage may be impeded, with waterlogging in furrows/hollows.

Community	Dominants	Distribution	Soils	Moisture
Ordinary damp meadows: Yorkshire-fog and soft-rush rush-pasture (MG10).	Creeping bent (*Agrostis stolonifera*), Yorkshire-fog, soft-rush (*Juncus effusus*) and creeping buttercup (*Ranunculus repens*).	Common from the lowlands into the upland fringe of England, in pastures and derelict farmland. This was perhaps the most common community in the drained grassland areas of the Fens.	Gleyed brown earths and alluvial soils – neutral to slightly acid. In more calcareous sites, hard-rush may replace the soft-rush.	Soils permanently damp due to ground or surface water.
Inundation grassland: red fescue, creeping bent and silverweed grassland (MG11).	Creeping bent, red fescue and silverweed (*Potentilla anserina*).	Probably frequent especially in west) in floodplains, upper saltmarsh and marginal habitats (i.e. by ditches, ponds and shores). Common on the silt fringe of the Fen basins.	Brown earths and alluvial soils (sand and shingle possess particular variants). Circumneutral but often brackish.	Soils moist to damp, but free-fdraining.
Upper saltmarsh ungrazed swards: tall fescue grassland (MG12).	Creeping bent, tall fescue (*Festuca arundinacea*) and red fescue.	Uncommon at the upper edge of saltmarshes and estuaries or on clay coastal cliffs. Likely to have been present on the coastal silt fringe.	Clays and silts.	Soils damp, but free-draining.
Inundation grassland: creeping bent and marsh foxtail grassland (MG13).	Creeping bent, marsh foxtail (*Alopecurus geniculatus*).	Widespread by water, especially in the East, on washland, floodplains and in seasonally wet hollows. Probably the most common community on river floodplain.	Circumneutral silts.	Soils damp, and sometimes waterlogged.

Definition of terms

1. The usage of the terms 'dry', 'moist', 'damp' and 'wet' follows that defined by the water indicator (F) values of Ellenberg (1988), i.e. occurrence in the gradient from dry shallow-soil, rocky slopes to marshy ground, and then from shallow to deep water.
2. 'Dominant' adapted from usage in the community description in the National Vegetation Classification (Rodwell 1991, 1992, 1995): i.e. species which occur in the majority of samples, but with ground cover in excess of 5 per cent and which also form a major part of the vegetation cover in at least some examples of that community.

Source: modified from Mountford, 1997

TABLE 14.2 Requirements of mire plant communities (M22)

Rich fen meadow

National Vegetation Classification	Dominant species	Occurrence (habitat)	Soil type	Watertable
Rich-fen meadows: blunt-flowered rush and marsh thistle fen-meadow (M22).	Marsh thistle (*Cirsium palustre*), marsh horsetail (*Equisetum palustris*), meadowsweet (*Filipendula ulmaria*), Yorkshire-fog (*Holcus lanatus*), blunt-flowered rush (*Juncus subnodulosus*), greater bird's-foot-trefoil (*Lotus pendunculatus*), water mint (*Mentha aquaticum*) and the moss *Calliergon cuspidatum*.	Occasional in central and eastern England, where rich fens have undergone some reclamation. More common the Fen basin than it is today.	On neutral to rather alkaline soils (pH 6–8) often peaty but also on alluvium and base-rich clays.	Soils moist to damp for most of the year, often due to flushes or springs.

Requirements of swamp communities (S5–7 and S22)

National Vegetation Classification	Dominant species	Occurrence (habitat)	Soil type	Watertable
Washland: reed sweet-grass swamp (S5).	Reed sweet-grass (*Glyceria maxima*).	Regularly inundated flood-plain wash-land (also common as a fringing swamp to still or sluggish eutrophic water).	Nutrient-rich, circumneutral or basic alluvium, or organic soil with regular inputs of mineral-rich water (pH>6.0).	Usually in waterlogged sites – with water at soil surface for most of the summer.
Tall sedge meadows: great pond-sedge swamp (S6).	Great pond-sedge (*Carex riparia*).	Wet hollows within flood meadows (also common dominant of swamp by still or sluggish, often eutrophic, water).	Mesotrophic to eutrophic, circumneutral mineral soils, more rarely on peat (pH 5.8–7.0).	Continuously waterlogged sites (up to 20cm of water).
Tall sedge meadows: lesser pond-sedge swamp (S7).	Lesser pond-sedge (*Carex acutiformis*).	Wet hollows within flood meadows (also a dominant of swamp by still or sluggish, often calcareous, water).	Moderately eutrophic, circumneutral to basic mineral soils (pH 6.0–6.8).	Continuously waterlogged sites (up to 20cm of water).
Floating sweet-grass hollows: floating sweet-grass water-margin vegetation (S22).	Floating sweet-grass (*Glyceria fluitans*).	Wet depressions in grassland, and along the margins of pools and drainage channels.	Mesotrophic to moderately eutrophic water on mineral, or more rarely, organic substrates (pH 5.0–7.0).	In grassland in waterlogged sites (community also occurs in up to 20cm of water).

Source: modified from Mountford, 1997

TABLE 14.3 Description, habitat conditions and range of fen communities

Community	Description	Habitat conditions and range
S24[1] Common reed-milk parsley fen (*Peucedano-Phragmitetum australis* and *Caricetum paniculatae peucedanotosum*).	Composed of tall monocotyledons (e.g. common reed and great fen sedge *cladium mariscus*) and herbaceous dicotyledons with lower layer of sedges and rushes and a patchy bryophyte layer. Generally species-rich (24).	Associated with flood-plain fens in England, especially in Broadland, where it occupies an intermediate zone between swamp and carr. pH, bicarbonate and calcium all moderate (pH 5.5–6.9). Mean water levels are low, though winter flooding occurs. Fertility is moderate; S24f is particularly low, whilst S24b is higher.
S25 Common reed-hemp agrimony fen (*Angelico-Phragmitetum australis*).	Characterised by tall monocotyledons and dicotyledons with variable amounts of small herbs and sedges. Less species-rich than S24 (11).	Found in flood-plain fens, open water transitions and 'sump' areas of valley mires in England and Wales. Generally associated with calcareous, base-rich water. Mean water table levels generally low, though higher than in S24.
S26 Common reed-common nettle fen (*Urtica dioica*).	Generally dominated by common reed and nettle but associates are variable. Generally species-poor (9).	Mainly associated with eutrophic, neutral to slightly basic water margins throughout the lowlands where winter flooding and summer drying occur.
S27 Bottle sedge-marsh cinquefoil fen (*Potentillo-Caricetum rostratae*).	Bottle sedge (*Carex rostrata*) may or may not be dominant, but marsh cinquefoil (*Potentilla palustris*) and bogbean (*Menyanthes trifoliata*) are constant. Species-poor (5). Potentially edges of meres	Almost exclusively a topogenous community in basin and flood-plain mires and may occur as a floating mat. Generally water levels are continuously high. pH, bicarbonate and calcium levels are low for rich fen communities. Fertility levels are high.[2] Absent from England south of East Anglia.

M5	Bottle sedge *Sphagnum squarrosum* mire.	Sedges and scattered poor fen herbs over a carpet of base-tolerant *Sphagnum*. Of medium species-rich (17).	May occur as a floating raft in fens which are mildly acid or moderatelt calcareous but oligotrophic. The two main habitats where it is found are in topen water transition and flood-plain fens. Water levels are usually high. Fairly local, mainly in the north-west of Britain.
M9[1]	Bottle sedge *Calliergon cuspidatum* mire (*Acrocladio-Caricetum diandrae* p.p. and *Peucedano-Phragmitetum caricetosum* p.p.).	Medium to tall fen vegetation, often species-rich, typically dominated by such species as bottle sedge, lesser tussock sedge (*Carex diandra*), slender sedge (*Carex lasiocarpa*), and common cotton grass. Sometimes there is patchy great fen sedge and/or common reed. Bryophytes, particularly *Calliergon* species, are conspicuous. Species-richness very variable (25).	In northern and western Britain mainly associated with basin fens, whilst in the South often hydroseral within flood-plain or even valley fens (but usually associated with topogenous hollows). Calcium and bicarbonate values are usually low and pH moderate. Mean water level is high. Low fertilities are associated with optimal community development.[2]
S1[1]	Tufted sedge swamp.	Vegetation dominated by tufted sedge (*Carex elata*) tussocks with some taller herbaceous dicotyledons. Generally species-poor (12). Found with S2 and S27. Probably present in the Fen basin.	Associated with open-water transitions, mesotrophic to eutrophic, shallow pools and turf-cuttings, only in west Norfolk, Cumbria and Anglesey. pH range 5.5–7.2 (Norfolk). Water levels up to +40cm.
S2[1]	Great fen sedge swamp and sedge-beds (*Cladietum marisci*).	Great fen sedge-dominated vegetation. Pure stands common and no other species frequent. Species-poor (7). A managed system for thatching.	Found in open-water transition, flood-plain and especially basin fens. Usually calcareous and base-rich. Shallow standing watertables. Tolerant of the range −15 to +40cm. Local, including Anglesey, Norfolk, Cheshire and Cumbria.

Community		Description	Habitat conditions and range
S3	Greater tussock sedge swamp (*Caricetum paniculatae typicum*).	Dominated by great tussock sedge tussocks. Species-poor (8). Associated with S4 and S13. Widespread in Fen basin but local.	Found in open-water transition, flood-plain and basin fens and in peat cuttings. Generally base-rich and calcareous (71–4 mg/1). pH range 7.1–8.1, mesotrophic to eutrophic. Able to tolerate a degree of seasonal watertable movement. Widespread but local.
S4	Common reed swamp and reed-beds.	Common reed (*Phragmites australis*) is the dominant. Generally species-poor (3), though variable, e.g. common marsh bedstraw (*Galium palustre*) sub-community is richer. Most common habitat.	Widespread in open-water transition and flood-plain fens, usually in hydroseral situations. Management extends the community into drier situations but water regimes can be variable. Does not have strict substrate preferences.
S5	Reed sweet grass swamp. (Reed sweet grass association).	Species-poor vegetation (4) dominated by reed sweet grass (*Glyceria maxima*) with a variable range of associates, e.g. great willow herb (*Epilobium hirsutum*), meadowsweet (*Filipendula ulmaria*), bittersweet (*Solanum dulcamera*). Common.	Mainly found in flood-plain fens (though not confined to fens), often on substrates containing a substantial mineral component, e.g. mineral alluvium. May develop as a floating raft. Mean values of pH, bicarbonate and calcium are high, though variable. Watertable levels have a low mean value. Associated with eutrophic, fertile conditions (particularly with high phosphate levels).[2] Widespread in the lowlands, but of very restricted occurrence in Wales.
S6	Great pond sedge swamp.	Large tufts of great pond sedge are dominant; hence stands are usually species-poor.	Characteristic of margins of standing or slow-moving water in mesotrophic to eutrophic conditions in the agricultural lowlands of England and Wales.

S7	Lesser pond sedge swamp.	Dominated by lesser pond sedge (Carex acutiformis). Common.	Eutrophic margins of slow-moving water.
S8	Common club rush swamp.	Typically with a somewhat open cover of common club rush (Schoenoplectus lacustris). Common.	Often occupies the deep-water limit of swamp vegetation in mesotrophic to eutrophic waters. Sub-communities are related to water depth and trophic status. Notably uncommon in Broadland.
S12	Bulrush swamp.	Bulrush (Typha latifolia) is dominant and stands are often species-poor (4). May be associated with S9 and grade landward into S25b. Common element at fringe of meres.	Widespread through the agricultural lowlands of England but less common in Wales and Scotland. Waters tend to be mesotrophic to eutrophic.
S13	Lesser bulrush swamp (lesser bulrush society).	Dominated by lesser bulrush (Typha angustifolia). Species-poor (4). May give way to S14 in shallower water. Common element at fringe of meres.	Found in standing or slow-moving water on silt, neutral to basic. Scattered distribution in England, becoming rare in Wales and to the North.
S14	Branched bur-reed swamp.	Branched bur-reed (Sparganium erectum) is generally dominant, but associates can be important. Common in ditches.	Very common in shallow mesotrophic to eutrophic water on a mineral substrate and found both in pools and alongside streams and rivers throughout the agricultural lowlands.
S15	Sweet flag swamp.	Sweet flag (Acorus calamus) may form an open or closed cover. Species-poor (6). Local.	Occurs in standing or slow-moving water 20–80cm deep. Substrate usually silt or clay. pH range 5.7–7.2. Scattered through the English lowlands.
S16	Arrow head swamp.	Arrow head (Sagittaria sagittifolia) is dominant and other species are usually only occasional. Common.	Most characteristic of moderately deep eutrophic waters and soft silty substrates. Water standing or slow-flowing. Scattered through the southern and central English lowlands.

Community	Description	Habitat conditions and range
S17 Cyperus sedge (*Carex pseudocyperus*) swamp.	Can form almost pure stands or be intermixed with other emergents. May be adjacent to S4 or associated with S24. Likely to be local.	Most typical of shallow, mesotrophic to eutrophic, standing or sluggish water. Patchily distributed in the English lowlands and most characteristic of the Midlands.
S18 False fox sedge	False fox sedge (*Carex otrubae*) forms a generally patchy cover and there can be a great variety of associates, e.g. soft rush and tall herbs, but most are not frequent. Edges of rivers.	Characteristic of clayey margins of standing or slow-moving, moderately eutrophic waters in the English and Welsh lowlands.
S19 Common spike rush swamp.	Dominated by common spike rush (*Eleocharis palustris*). Generally species-poor (7). Locally frequent.	Found in a wide variety of sites, often over silt, in mesotrophic to eutrophic, standing or running water throughout Britain.
S20 Grey club rush swamp.	Grey club rush (*Schoenoplectus tabernaemontani*) dominates, with various species of saltmarsh and of disturbed and/or moist soils.	Found most frequently in moist, brackish sites with soft gleys of silt or clay.
S22 Floating sweetgrass swamp.	Floating sweet grass (*Glyceria fluitans*) occurs as low mat or floating carpet. Generally species-poor (5). Common in ditch systems.	Characteristic of shallow, standing or slow-moving water on a mineral substrate in the agricultural lowlands.
M10 *Dioecious* sedge-common butterwort mire.	In general, this is a low-growing small-sedge community. Bog rush and purple moor grass may be present. Moderate to high species-richness (25). May be associated with a wide variety of peripheral communities. Local at the edge of the Fen basin often near springs.	Mainly occurs in small, often isolated, spring fens, though larger stands occur if springs amalgamate to form a flushed slope. Occurs on a wide range of soils, usually not peaty. Bedrock is often limestone. pH and calcium levels are high (similar to M13). Water levels moderate, redox high. Fertility values are low.[2]

M13 Black bogrush-blunt-flowered rush mire. — Dominated by black bog rush (*Schoenus nigricans*) and blunt-flowered rush (*Juncus subnodulosus*) but with a wide range of low-growing associates. Species-rich (27). — Predominantly found in valley and spring fens, on a wide variety of soils, but usually associated with base-rich water, pH 6.5–8.0. Summer water levels range from low to high.

Notes

Community names in brackets in the first column are those of Dr B. D. Wheeler.
The figures in brackets in the second column represent species per 2 x 2m quadrat.
1. Rare or highly localised communities or sub-communities.
2. Information presented here is taken from Wheeler and Shaw (1987).

Approximate ranges of environmental variables in five descriptive categories	Very low	Low	Moderate	High	Very high
pH (water)	<5.0	5.0–6.0	6.0–6.6	6.7–7.1	>7.1
Bicarbonate (mg/1) (water)	<105	106–250	251–369	370–460	>460
Calcium (mg/1) (peat)	<620	621–1,200	1,200–2,000	2,000–3,000	>3,000
Water depth (cm)	<-25	-25 – -10	-9 – -1	+1–+9	>+9

On wet and poorly drained land edging true fen habitats the blunt flowered rush, marsh thistles and the Yorkshire-fog grass typify what is known as 'rich fen meadow' (Table 14.2). Marsh orchids would here stain the meadow pink in spring by the abundance of flowering spikes. Rich fen meadow is the consequence of attempts made to drain the drier edges of true fen habitat. This habitat was managed for reed and different species of sedge. Reed and sedge were used for thatching cottage roofs; saw sedge or great fen sedge was used to ridge the thatch; other sedges were used as bedding for cattle; and the sweet flag was used to perfume the cattle byres and stables. Where the fen was cut for peat, to provide fuel, a range of fen plants would grow on the varying degrees of wetness and dryness resulting fom the uneven pattern of cutting and terrain. Plants such as yellow loosestrife or marsh pea would grow on the drier ridges. The rarest of these plants would have been the fen orchid, which would have been common by today's standard as this plant is now a national rarity (found only in fifteen or less 10 x 10km squares throughout the British Isles). *Sphagnum* mosses would have colonised the wetter cuts, and the insectivorous sundews would have been commonplace. Insectivorous butterworts would be found on spring lines.

Parts of the drier fen would have been managed to grow alder. This would have been coppiced and used to produce staves, wattle fencing and wattle sides for the cottages. Within the alder carr (wet woodland) and along its edges, ferns – such as royal fern, marsh fern and crested buckler fern – would have grown. These are now relatively rare in East Anglia as a whole, their population centre being in the Norfolk Broads. The delicate white flowering spikes of wintergreen, and the powdery purple of the marsh helleborine, would have been found in clearings or in the edges of the carr woodland.

In the more acid areas of a fen the bush, bog myrtle, would have perhaps foreshadowed the development of acid peats where *Sphagnum* mosses would have formed russet and yellow cushions of wetness. At the edges of such damp areas would be found heather, and perhaps amongst it the white flowers of the grass of parnassus.

A rare habitat within the Fens would have been that of true raised mire or bog. Only one example is recorded with any certainty, that of Holme Fen, Huntingdonshire. This was formed by a quirk of hydrology: the impeded drainage of streams entering Whittlesey Mere and Trundle Mere allowed peat to develop in the island(s) between the streams. Layer upon layer of *Sphagnum* grew, each plant 'sucking in' and retaining its own water, even though the bog was raised above the level of the surrounding fens.

Ditch, sluggish river, stream and mere would have contained a rich community of pondweeds, white and yellow water lily, water milfoil, water crowfoot, bladderworts, mares tail, water violets, duckweeds, frogbit, lesser water plantain, even perhaps floating leaved water plantain or least bur-reed. In trophic terms these meres would be upper mesotrophic to naturally eutrophic. The lake or mere community containing frogbit and a variety of pondweeds is now only found in a lake on Anglesey, North Wales and in Upper Lough Erne in Northern Ireland, such as been their demise through habitat loss and pollution. However, this community is still widespread in the least polluted and sympathetically managed ditches of southern and eastern England.

Around the edges of the meres, bulrush, water plantain, reed and sedge would have formed a swamp community with alder carr in the drier areas.

— Animal life in the Fens —

With an abundance and variety of habitat the Fens would have been vibrant and literally buzzing with life. The myriad of insect life, including the malarial mosquito (finally eliminated with the disease around 1920) would have been preyed on in their aquatic or larval stage by fish and birds. Once airborne, they were prey to dragonfly and birds such as reed, sedge or marsh warbler, bearded tit, swallow or sand martin. Hobbies would feed on the dragonflies using the carr woodland as a nesting area. Sparrowhawk would forage on the small birds, and buzzards feeding on young rabbits would soar in the thermals generated by the 'scarp' slopes or relatively high ground edging the fen area. Red kites would scavenge the dead rabbits or other carrion.

Butterflies would have been far more numerous than they are today, and the wooded fringes of the Fens, containing blackthorn, would have been home to now rare species, such as the black hairstreak. The swallowtail butterfly has been recorded at Wicken Fen: it is difficult to say whether it was ever common in the Fens, since its food plant, milk parsley, seems to have been ousted at an early date by the Cambridge milk parsley, which was found in several fens in the area, although now survives only at Chippenham Fen NNR. Certainly, the dazzling brilliance of the copper colour on the large copper butterfly, extinct in Britain since 1910, would have been a common sight in the true fens scattered throughout the area.

At the edges of the meres, coot, moorhen, little grebe, mallard and teal would have been found. Skulking amongst the reeds there would be water rail and spotted crake. In the reeds, bittern would boom; in the wet grassland corncrake would rasp out their monotonous 'crek crek', and snipe would drum. Redshank, godwit and lapwing would utter their plaintive calls. Heron and otter would feed on the abundance of fish. Autumn would have seen a host of wildfowl: records of sales from Ramsey market suggest that that their numbers must have been in the hundreds of thousands. As autumn progressed the Bewick and whooper swans would have arrived from Siberia, just as they do today. Bean, pink foot and grey-lag geese would have all been shot by the fen men in their long boats containing punt guns, capable of bringing down twenty or more wildfowl in one shot.

The Fens were a true wetland, and supported a sustainable economy. The wealth of plant and animal life would have been truly staggering. But, as we have seen, new forms of land use finally prevailed, and the Fens were largely destroyed. 'Progress' and 'improvement' won the day, and even in 1967 Dring was able to describe:

> Until the 17th century it was a waste of peat bogs interspersed with islands, sparsely inhabited by an independent race of people living in primitive conditions and existing mainly on the fish and fowl they obtained from the rivers and meres. The foresight and determination of a few men changed this into what is now one of the most highly productive areas in the country. Land, once valueless, is now worth over £400 an acre and it produces rich yields of sugar beet, potatoes, celery, carrots, grain and other crops which are supplied to the markets of London and elsewhere. (Dring, 1967.)

CHAPTER 15

Water management systems: drainage and conservation

JOHN SHEAIL

— INTRODUCTION —

Alan Bloom, in his account of wartime farming in the East Anglian Fen, recalled struggling to remove the remains of a large bog-oak. This was buried in a field being reclaimed in order to help relieve the desperate food shortage. A passer-by stopped, recalled his days as a Cambridge student botanising in the locality, and looked forward to the day when the land would again be abandoned. It would once again becoming a 'buffer' to the adjacent nature reserve owned by the National Trust. As Bloom (1944) wrote, 'we were in opposite camps, and he could no more appreciate my line of reasoning than I could see his point of view'.

The anecdote is cited by Jeremy Purseglove in his *Taming the Flood* to illustrate how 'this conflict of values in which neither side has an absolute monopoly of truth has rumbled on'. To Purseglove (1988, pp. 73–5), it assumed the dimensions of a battle which could be traced through the post-Second World War years. Locations included the East Anglian Broadland, the Somerset Levels and Moors, and several other wetland areas of the UK.

The purpose of this concluding chapter is to look more closely at how that dichotomy of perception (conservation versus production) came to impinge so largely on decisions about the future management of the countryside, and especially of 'wetland' areas. It is contended that both the drainage industry and conservation movement were imbued with a sense of historical mission. For example, a leaflet published by the ADAS Land Drainage Service (no date) of the Ministry of Agriculture, Fisheries and Food celebrated the 2,000-year history of agricultural drainage, beginning with the Roman draining-spade and today advanced through 'trenchless drainage with "filter wrapped" plastic pipes and laser grading'.

In their separate ways, both the agricultural industry and conservationists saw the post-war years as an exceptional opportunity. For the land-drainage industry, historical experience suggested that the unprecedented levels of political support and material investment in drainage works (shown by grant aid for underdrainage) would be short-lived. Compared with farming and forestry, the conservation bodies recognised that, as a third force in the countryside, their resources were meagre. But unless wildlife habitat was protected with the utmost expedition, the last examples of the scenery and wildlife of semi-natural ecosystems would be degraded, if not destroyed once and for all. As an exemplar of the large-scale changes brought about in the use and management of the countryside, drainage schemes became the stuff of polemics.

— BEFORE 1930 —

There are two crucial dates in the evolving relationship. The year 1949 marks the beginning of the modern nature-conservation movement. In that year, the first official bodies (the National Parks Commission and Nature Conservancy) were appointed and their powers conferred under the National Parks and Access to the Countryside Act. The year 1930 has a similarly practical, and indeed symbolic, significance for the drainage industry: it was the year in which the Land Drainage Act was passed.

This act had long antecedents. The Bill of Sewers of 1531, the first of some 16 public statutes, had laid down criteria for appointing the oldest of the three forms of surviving drainage authority, the Commissions of Sewers. By 1930, there were 49 such bodies. A further 198 authorities had been established under private acts. The third group comprised 114 elective Drainage Boards created under enclosure awards and, since the Land Drainage Act of 1861 (24 & 25 George V, c. 133), by the Board of Agriculture (which became a ministry in 1918). The establishment of boards was severely constrained both by the depressed state of arable farming from the late 1870s and by 'the somewhat cumbersome proceedings' that had to be followed. The consent of owners of two-thirds of the land was required before a Draft Order could be confirmed. In order to accelerate the process, a further Land Drainage Act of 1918 (8 & 9 George V, c. 17) empowered not only landowners, but also County Councils, to petition the

Ministry to form, merge or alter the boundaries of a board, or indeed transfer such powers to a council. An Order might proceed if proprietors of no more than one-third of the land objected (Roseveare, 1932).

The unprecedented concern for food production during the First World War, and the resources consequently invested, greatly boosted drainage activity. Using wartime powers, the Board of Agriculture typically required the West Riding of Yorkshire County War Agricultural Executive Committee to undertake drainage works. Priority was given to existing courses, removing obstructions, easing sharp bends, and repairing floodbanks. The improvements made to some 110km of channel in West Yorkshire enabled several thousand acres to be brought into cultivation, often for the first time in thirty or forty years. A further 30km of neglected and obstructed drains were improved during the winters of 1921-2 and 1922-3, by the County Agricultural Committee, acting as agent to the Ministry under the Government's Unemployment Relief Works schemes.

Such intervention by central Government required Exchequer grants of up to 75 per cent. It recalled (but on a smaller scale) intervention by the Board of Agriculture during the Napoleonic Wars. Not only did these early twentieth-century projects demonstrate what could be achieved through co-ordinated drainage effort; they also highlighted the absence of measures to prevent such works reverting to their former condition. Whilst there might be 'a multiplicity of drainage authorities' on paper, most were moribund or had never properly functioned. For example, although twenty-two enclosure awards in the West Riding made provisions for drainage, they were disregarded in three-quarters of the cases. As the Chief Technical Adviser of the wartime Food Production Department later remarked, it seemed unthinkable that 'such drainage and tidal defence works should only be maintained under conditions and during periods of national misfortune' (West Yorkshire Record Office, RD1/5/9, Boxes 1,6,7,13 and 28). The West Riding County Council followed the precedent set by Lancashire in applying for the powers of a drainage authority for the entire administrative county. The Council's Agricultural Committee would have general oversight of existing bodies, and provide the necessary organisational structure where required in other areas. The West Riding of Yorkshire County Council Bill received the Royal Assent in August 1923 (13 & 14 George V, c. xcviii).

Whereas the greater part of the Yorkshire Ouse watershed fell within the administrative county of the West Riding, the catchment of the more southerly Bedfordshire Ouse that drained into the Wash, 800,000ha in all, included parts of eleven counties. Under wartime powers, the Board of Agriculture appointed a Lower Ouse Drainage Board, whose proposed scheme to improve the tidal portion of the river and construct training walls into the Wash won general approval from technical witnesses at an inquiry in 1917. A draft Order under the Act of 1918 proposed that the scheme should be implemented under an Ouse Drainage Board. This would cover both the fens, and those parts of the higher watershed within a line drawn 8ft (2.44m) above the level of the highest flood. In a parliamentary debate to confirm what was, in effect, the first 'really large systematic attempt to drain a big area', Members of Parliament protested that those living some 60 or 70 miles (95–115km) upriver should be required to contribute to the Board's rateable income. For the most part, flooding was welcomed in the upper reaches for its fertilising effect on adjacent grasslands. Members of 'lowland' constituencies supported the Parliamentary Secretary of the Ministry of Agriculture, Arthur Boscawen, in his insistence that the 'uplands' should, nevertheless, take some responsibility for the discharge of their waters to the sea (PD, Commons, 127: 1929–51).

Although the enactment of the Order in October 1920 concentrated responsibility for the main channels and outfall, and made provision for meeting the costs of that work, there quickly developed 'an acute financial situation'. Not only was it an 'extremely onerous, expensive and lengthy business' to prepare the rate books, but there was acute disagreement among members of the Drainage Board. Most importantly, there was passive resistance among ratepayers, first in the 'uplands' and then more generally. Having met a deputation, the Minister drafted an amending Order in 1923, which effectively put the 'uplands' under separate administration. The only point of consensus to emerge from the consequent eight-day inquiry, in the spring of 1924, was the need for further legislation and substantial Exchequer support towards the improvement works, estimated at £2.5 million. The Ministry offered, in July, to meet the full cost, provided two-thirds were repaid over a 30-year period by means of a flat rate levied over the whole District, excluding the 'uplands'. The Drainage Board was in no position to accept the offer. The banks had set a limit of £70,000 on its borrowings, namely the anticipated rateable income. Some £40,000 remained to be collected. The Ministry concluded that the only course was to appoint a special Commission to investigate 'the whole problem of the River Ouse' (PRO, MAF 49, 311–13 & 316–24).

A small expert Commission found that, without the engineering scheme, inundation was inevitable. Its report of December 1925 warned that there was a real danger of the whole District 'returning to its original condition of swamp'. There was so intimate a connection between the works required to remove the 'uplands' discharge, to prevent the accumulation of silt in the tidal channel, and to stabilise the banks that nothing but the completion of the entire scheme would suffice. That could only be achieved if the basis of rating was broadened. Although ambiguous, the statute of 1531 had been construed as limiting liability to those persons who benefited directly from such works and the avoidance of danger: the jurisdiction of a drainage board was thus usually interpreted as being confined to land subject to flooding. The Commission believed the entire catchment area should be treated as 'a single community which cannot be artificially divided'. Although the main burden of charge should continue to fall on those areas directly to benefit, it was reasonable to expect a contribution from the rest of the watershed as an acknowledgement of community interest. Half the capital and revenue expenditure of the Ouse Drainage Board should be provided by a precept on the internal drainage boards and, as a new principle, by the local authorities. The remaining cost should be met by the Exchequer. Not only did the outfall present exceptional difficulties, but the works were the only means of safeguarding an extensive area of fertile agricultural land from catastrophe (PRO, MAF 49, 325-9; Ministry of Agriculture, 1926).

The Minister, Walter Guinness, moved a Bill, in June 1927, to implement the Commission's findings. It was a hybrid Bill, in the sense that it was a Government measure that anticipated the expenditure of public moneys, yet dealt so closely with private interests as to require scrutiny in the manner of a Local or Private Bill. The Member for south-west Norfolk found it remarkable for uniting in opposition each of the four classes principally affected, namely the ratepayers of the 'lowlands' and 'uplands', those who lived in a relevant county but not in the watershed, and fourthly the taxpayer. In his judgement, much of the opposition arose from the Government's failure to acknowledge the Commission's assumption that, if accepted, its findings would be applied to the whole country (PD, Commons, 207:1209-86). They were in fact only adopted in respect of the Ouse. A further Royal Commission on Land Drainage was appointed in March 1927 to go into the wider question. Having heard through legal counsel evidence from a succession of witnesses, a joint Committee of both Houses of Parliament decided by a majority of five to three that the Ouse Drainage Bill should not proceed. Although the Lord Chairman followed the custom of not giving reasons, it was assumed the majority thought such a far-reaching decision must await the findings of the Royal Commission (Joint Select Committee, 1927; Roseveare, 1932, p. 181).

— The Land Drainage Act of 1930 —

In evidence to the Royal Commission, the Ministry's Chief Drainage Engineer calculated that 1.75 million acres (709,000ha) of land in England and Wales were in 'urgent need of drainage' (Roseveare, 1932, p. 182). As anticipated, the Commission's report of December 1927 recommended that general drainage law should be consolidated and considerably amended. Although the principle of benefit and escapement from danger should be retained in determining how the internal drainage bodies were elected and their costs met, new drainage authorities should be established for the larger catchment areas, with comprehensive powers to undertake works and raise the necessary revenue from the whole watershed (Royal Commission, 1927).

A Land Drainage Bill based closely on the Royal Commission's findings received the Royal Assent in August 1930 (20 & 21 George V, c. 44). Forty-seven Catchment Boards were to be appointed in England and Wales by November 1931 – they included both the Yorkshire Ouse Catchment Board and more southerly Bedfordshire Ouse Catchment Board. Each might comprise up to thirty-one members, of whom two-thirds would represent the relevant County Councils and County Borough Councils. Besides one member representing himself, the Minister was to appoint the remainder as representatives of the lands that would directly benefit from the drainage works. As well as general oversight of its watershed, each Board would have direct responsibility for those watercourses identified by the Minister as 'main rivers'. Further local drainage bodies, known as Internal Drainage Boards (IDBs), would be established where they might help avoid danger or confer significant benefit to the land. The income of the Catchment Boards would be made up of three elements, namely the levy of lump-sum precepts on the IDBs and County (and County Borough) Councils. Thirdly, and most crucially in securing the consent of the various interests to the Bill, the Minister was empowered to make grants towards new works and the improvement of existing works.

As the Ministry's Chief Drainage Engineer remarked, one could not fail, in travelling about the country, to be impressed by the deterioration in drainage conditions over the previous five or six years, much of it due to the neglect of small ditches and streams (Roseveare, 1932, p. 182). Farmers simply could not afford such work in hard times. In his official history of agriculture in the Second World War, Murray (1955, p. 53) described how, between the wars, field drains and ditches had become obstructed, and adjacent lands waterlogged; fences and hedges went unrepaired, and grasslands under-stocked. It was fear of a further war that caused the Cabinet to appoint, in February 1937, an inter-departmental committee 'to make definite proposals for increasing the productivity of our soil', and Parliament to enact an Agriculture Bill later that year, whereby IDBs and local authorities could also apply for Exchequer grants towards the improvement of minor watercourses (1 Edward VIII & 1 George VI, c. 70). By June 1940, field drainage had become one of the most obvious ways of 'effecting a considerable and immediate improvement in regard to food production'. Catchment Boards might now apply not only for grants of 15–75 per cent for work on the principal rivers, but for 50 per cent grants towards improvements to minor watercourses and ditches outside internal drainage districts. Grants of 50 per cent were offered to individual owners and occupiers for ditching, mole-drainage, and tile-drainage, where approved by the County War Agricultural Executive Committees (CWAECs) (Agriculture (Miscellaneous War Provisions) Act, 1940, 3 & 4 George VI, c. 14). As many as 40,000 farm-drainage schemes were approved in England and Wales, covering 1.214 million acres (500,000ha) in 1941. A further 1.9 million acres (770,000ha) were covered in the ensuing two years (Murray, 1955, pp. 128–9).

The files used by officials of the Ministry of Agriculture reveal much frustration and disappointment as to the slow and patchy uptake of drainage schemes during the post-war years. Resolutions were soon received from the CWAECs, Land Agents' Society and National Farmers' Union (NFU), claiming that the scope for carrying out field-drainage works had been exhausted within the limits set for grant aid. Although the Ministry had discretionary powers to award higher grants, officials feared their use would 'rule out the incentive to do the job economically'. There was particular resentment among farmers in the English northern counties that holdings only 'a few miles away over the Border' should be free of any upper limit under separate Scottish legislation. The Ministry insisted that there were bound to be differences in policy, reflecting 'basic climatic and geographical dissimilarities'. It was commonplace in Scotland to come across 'wet spots' where drainage was extremely expensive if measured 'on a strictly per-acre basis', but much more reasonable where assessed 'from the standing of the farm economy as a whole'. Rather than raise or abolish limits to grant aid in an attempt to secure greater equity between holdings of different locations or size, Ministry officials argued the more economical and practical course of finding ways of reducing costs. There was an enormous incentive for introducing new drainage techniques, particularly through making greater use of machinery (PRO, MAF 49, 836-7, 888-93 & 1831, and MAF 218, 12).

The trend towards more comprehensive governance of watersheds was sustained by the River Boards Act of 1948 (11 & 12 George VI, c. 32), which brought together under newly-appointed River Boards the powers previously exercised in England and Wales by the Catchment Boards, Fisheries Boards, and local authorities in respect of pollution control. In the twenty years since the Act of 1930, some £20 million had been spent by the Catchment Boards on new and improved drainage works, and a further £10 million on internal drainage systems under the Agriculture Act of 1937. Some £23 million had been invested since 1940 on field-drainage works. A large proportion of hedge-side ditches had been re-excavated and underdrainage systems renovated. Whilst acknowledging the achievement, commentators emphasised how much remained to be done before the general state of drainage matched that attained in the previous period of agricultural prosperity in the nineteenth century (Johnson, 1954, p. 628).

A paper given by the Ministry's Chief Drainage Engineer, E. A. G. Johnson, to the Institution of Civil Engineers in 1954 covered both the purpose and execution of drainage schemes. As he emphasised, the goal was not one of preventing flooding altogether – even if this was a realistic option. However important what he called 'the amenity value of the rivers and the consequent need to keep them tidy', the focus of attention was not the state of the river channel and its capacity to accommodate floodwaters, but what were the consequences of that flooding, and what were the benefits to be derived from its prevention, say in terms of food production. Whilst it might be fairly easy to relate the expense of replacing an aged pumping-plant to the cost of damage caused by its failure, the economics became more theoretical where a major scheme was contemplated. If based solely on increased land or rental values, few such arterial schemes could be regarded as sound. It was one of the roles of the River Boards (after 1948) and Exchequer grant aid to compensate for what Johnson (1954, pp. 604-6) called 'the possible lack of direct or immediate economic relationship'. It called for the closest understanding on the part of the drainage engineer of the agriculturalist's longer-term needs.

— THE OPTIMAL LIMITS TO DRAINAGE —

There now seemed some point in debating the optimal limits of drainage schemes, and in defining the most appropriate form that their improvement and management should take (Trafford, 1970). Section 96 of the Agriculture Act of 1947 (10 & 11 George VI, c. 48) had placed on a peacetime footing many of the statutory instruments required for the continued institutional support of farming. Continued recognition of the need for agricultural-price support was made conditional on increasing efficiency on the part of the farmer. In moving a further Agriculture Bill, in 1957, the Minister asserted it was emphatically 'not a Measure to enable less-successful farmers to stay in business'. Far less was it a Bill to perpetuate 'the existing pattern of production'. Its goal was 'the long-term economic efficiency and competitiveness of the industry' (PD, Commons, 567, 807–14). Such exhortations made conflict ever more likely with another set of rural interests enjoying unprecedented support from the State.

A Nature Conservancy had been appointed, in 1949, charged with the responsibility of establishing and managing a series of National Nature Reserves (NNRs), giving advice on nature conservation generally, and undertaking relevant research. It was a hybrid body in the sense of being a research council, yet deriving its statutory powers from the National Parks and Access to the Countryside Act (which otherwise related to the planning sector of Government). On the premise that NNRs could never encompass the entire range of 'changing natural conditions', the Conservancy was required by the Act to notify local planning authorities of those 'Areas of Special Scientific Interest (SSSIs)' which were not currently held and managed as NNRs. The authorities were required to consult the Conservancy before granting a statutory planning consent for building and other forms of 'development'. The rapidity and scale of change in the use of the countryside soon caused the SSSI to assume a significance never intended (Sheail, 1995).

Besides drawing up management plans for NNRs, the Conservancy's regional officers were to undertake surveys of the wildlife resource of the wider countryside. A memorandum of November 1963 by the Regional Officer for East Anglia, Bruce Forman, cited how the Leziate SSSI in west Norfolk had been completely destroyed, and that of the adjoining Derby and Sugar Fens SSSI altered and much reduced. This was a result of drainage works begun in 1951, with the support of a retrospective grant in 1954, and completed in 1958, following a further grant. Almost the entire Nene Washes had been cultivated after the recent improvement of water control in the adjacent ditches due to downstream deepening of the main drainage channels. The Huntingdonshire NNRs of Holme Fen and Woodwalton Fen survived as 'islands' of relict fen within 'a vast expanse of highly developed agricultural land'. At the Holme Fen NNR, and at those of Calthorpe Broad and Redgrave and Lopham Fens in Broadland, where there had previously been little management of the sites, the river boards and IDBs strenuously opposed moves to reinstate a higher watertable. The Conservancy faced the choice of either abandoning the conservation interest, or investing heavily in impermeable banks and water-pumping equipment, so as to divorce the drainage regime of the reserves from the surrounding land (PRO, FT 3, 107).

As Forman recounted, a number of factors exacerbated the problem. Modern machinery had become so powerful as to make possible what, even a few years before, would have been considered impossible. In the sense that drainage of one area often made that of another easier, the effect was cumulative. A further factor was 'the tremendous increase in recent years in the numbers of IDBs', which saw their sole function as that of organising farmers and pooling their resources for drainage works. Not only was it a powerful means of forcing the 'less ambitious and "progressive" farmers' into line but, in Forman's words, 'any sympathetic supporter of conservation finds himself in hostile company if his sympathies show the slightest sign of conflicting with the general aims of the IDB'. With their 'basic lack of ecological knowledge and very limited outlook', the only yardstick by which farmers or the IDBs, River Boards and the Ministry could measure the value of an area was by its agricultural productivity. With only the vaguest idea of the purpose of the Nature Conservancy, they regarded its work as 'sentimental humbug', only to be tolerated where it did not interfere with farming. Even in the rare case of 'a partly sympathetic official', it invariably happened that his hands were tied. Any mild influence that he might try to exert counted for nothing once a farmer's appetite for reclamation was whetted. It was rarely worth bargaining for compromise. Any concession of land was invariably too small to provide a viable unit.

In Forman's judgement, naturalists would achieve little for as long as they depended on owners acting 'out of the goodness of their hearts'. The Nature Conservancy would be in a considerably stronger position if it could pay for the privileges sought. The question of compensation struck, however, at the root of the problem. The costs of such management agreements could only be met by a willingness on the part of Government to meet the almost open-ended costs. Forman might have identified a further obstacle. What drainage interests might perceive as privileges, conservationists might equally regard as their right, on behalf of society at large, to demand. From that perspective, there was a strong case for extending those aspects of statutory town and country planning that imposed constraint on development without liability for recompense, to cover such changes in agricultural use and management as land drainage. That course would be cheaper for the Exchequer, but even more likely to offend powerful networks of vested interests.

— Amberley Wild Brooks —

As with more 'progressive' drainage interests between the two world wars, the conservation movement drew considerable strength from a series of well-publicised clashes in determining the management of specific catchments (Purseglove, 1988). The powers exercised by the River Boards were first subsumed into River Authorities in 1964, and then into Water Authorities in 1974. The Southern Water Authority (SWA) applied, in July 1977, for grant aid towards a pumped drainage scheme for Amberley Wild Brooks, a pastureland of about 365ha within the floodplain of the river Arun in West Sussex. The Authority sought an Exchequer grant of 72 per cent of the total cost of £339,000. There was strenuous opposition, both locally and from national conservation bodies, concerning the impact of the proposed scheme on this secondary wetland. The Minister of Agriculture acceded to representations from the SWA, the Countryside Commission and the Nature Conservancy Council (the successor body to the Nature Conservancy) for a public inquiry. It was held in March 1978 (Public Local Inquiry, 1978).

The intention of the witnesses assembled by the Regional Land Drainage Committee of the SWA was to demonstrate how the proposals represented the inevitable next step in the long-considered and incremental improvement of drainage and, therefore, agricultural productivity. The flooding during the (notably wet) winter of 1974–5 had brought matters to a head. Whilst the Water Act of 1973 had enjoined the Water Authorities and Minister to have regard to the impact of any proposal on the appearance, wildlife and historical features of the countryside, the same measure also required them to identify future drainage requirements. The Country Landowners' Association (CLA) and National Farmers' Union (NFU) had demonstrated their concern to protect amenity and wildlife in a joint publication, *Caring for the Countryside* (CLA et al. 1984), but they too pointed to the recent White Paper of the Agricultural Departments, *Food from our Resources*, as making clear that priority should still be given to increasing agricultural productivity (Secretary of State for Northern Ireland et al., 1976). The proposed scheme would raise the agricultural grading of the Wild Brooks to the better Grade III and, in some parts, Grade IV.

If the intention of the Public Inquiry was to demonstrate how the drainage industry was providing essential support to progressive farming interests, the precedent was not a happy one. The Minister supported his Inspector in concluding that insufficient attention had been given to the economic, let alone conservation, aspect. Although none challenged the efficacy of the proposed pumping installation in upgrading the land's economic potential, a large number of deficiencies was identified on the part of the SWA. Besides being subjective in its assessment of farmers' intentions, the Authority had taken no account of the investment required to capitalise on the drainage scheme (Bowers, 1983). On the basis of a cost-benefit ratio that might range from 0.95 to 1.06, calculated by the expert (drainage) assessor appointed to the enquiry (from the Harper-Adams Agricultural College), grant aid could not possibly be justified. The Inspector similarly agreed with his expert assessor on the conservation side (from the Department of Biological Sciences at Exeter).

The probable agricultural benefit could not possibly compensate for the loss of wildlife and amenity value. Such wetlands occupied only 2 per cent of west Sussex. As the Inspector's report noted, floristic richness was not to be judged solely in terms of abundance or rarities, important though they were, but rather as 'the summation in which common, local and rare species grew together and exploited the available habitats'. Whilst present farm practices were largely sympathetic to the retention of 'the natural balance', the scheme would lead to 'substantial impoverishment of species and destruction of the unique peat-deposits'.

In emphasising the relict nature and irreplaceability of such wet-grazing marsh-systems, witnesses for the Nature Conservancy Council and Sussex Trust for Nature Conservation spoke of the need for positive action to conserve the last examples of a once common habitat. Once lost, they could never be reinstated. Such instances of conflict further substantiated the demands made in the annual report of the NCC for 1976-7. Besides safeguarding those 'living' relics, there had to be more explicit provision for nature conservation in the wider use of the countryside. Within a more consciously planned land-use strategy, nature conservation should be consciously promoted, with food production and other requirements, as 'vital elements in the wealth and heritage of the nation' (Nature Conservancy Council, 1977). Conservation as an 'environment good' had now fully emerged.

— Agriculture and the environment —

The Public Inquiry into Amberley Wild Brooks provided not only a critique of many of the assumptions behind the Water Act of 1973 but was itself part of the 'warming up' for what became the fierce controversy surrounding the Wildlife and Countryside Act of 1981. Whilst the Department of the Environment had intended the Water Act as a radical advance in integrated catchment management, the more incremental approach favoured by the Ministry of Agriculture on behalf of drainage interests had the effect of reinforcing the polarisation of positions. Through the tactic of threatening to disrupt the parliamentary timetable, those MPs sympathetic to drainage interests further ensured that the regional and local drainage committees appointed by the new Water Authorities enjoyed considerable autonomy. Where the overriding purpose of the Act was to organise water-resource management on a functional basis, land drainage continued to operate through the local IDBs, the drainage water authorities and the Ministry on a client basis (Richardson et al., 1978). In Purseglove's words, land drainage remained 'the Vatican of the water industry: a state within a state' (Purseglove, 1988, p. 214). Politically so strong, these interests resisted all moves to extend the scope of statutory planning to cover agricultural changes in land use. They were well able to drive hard bargains under the Wildlife and Countryside Act. There were bound to be heavy demands for compensation to cover the income foregone, where the NCC negotiated a management agreement to prevent a 'potentially damaging operation' on a SSSI.

In its scrutiny of the working of the 1981 Act, the House of Commons' Environment Committee affirmed the wisdom of the voluntary approach. The regulation of agricultural operations through an extension of town and country planning would never have secured the positive management inputs required for protecting amenity and wildlife. The Committee recognised, however, that land drainage was responsible for some of the most striking changes in landscape and 'the ecology of a wildlife habitat'. Its members were much moved by a visit, in November 1984, to the Swale SSSI of over 570ha in the Isle of Sheppey. Standing on the old sea wall, the Committee saw on one side a Royal Society for the Protection of Birds (RSPB) reserve of rich marshland, partly re-flooded and criss-crossed with a network of dykes. There were huge flocks of birds, mainly duck, geese and waders. On the other side was a tenant farm. Although most was marsh, it was more barren, with only very sparse patches of standing water. There were nevertheless many species of both fauna and flora. In the distance, there was a drained and cultivated area that, in contrast to 'the bleak, wild landscape of the marshes', comprised fenced fields of winter wheat and oil-seed rape. Without ponds, dykes or fleets, the only birds were crows, pigeons and other common agricultural species. There were certainly no 'outstanding assemblages' of rare species. Its bleakness was of another kind – tidy, domestic and sterile (House of Commons, 1985).

To the Committee, the mosaic of conditions encountered in that part of the north Kent coast highlighted the need to redirect agricultural grant aid. The tenant had originally submitted a proposal, in 1982, to underdrain and convert the entire marsh of 1,000 acres (400ha) over a three-year period. The NCC had offered to negotiate a management agreement. The Committee took heart from the tenant's rejection of the largely negative conditions, claiming that, if agricultural improvement was precluded, he wanted 'an active co-operating role'. Such instances of landowners and farmers, and their representative bodies, wanting more conservation in agricultural practice caused the Committee to press even more strongly for the Ministry to find ways of adapting the structures of the EEC Common Agricultural Policy so as to provide the resources for such positive acts of management as re-flooding, controlled grazing, and say the creation of paths and viewing points for visitors.

Whilst asserting the UK Government could not act in isolation from the remainder of Europe, the Ministry undertook to press for powers that would 'enable us in environmentally sensitive areas to encourage farming practices which are consonant with conservation'. Through perseverance, UK Ministers secured under Article 19 of the EEC Regulation 979/85 powers whereby Member States could make payments to farmers for the provision of 'public goods'. It did more than enable the Halvergate experiment in the Norfolk Broads to be extended. Where the environment sector of Government was responsible for existing designations, the Agriculture Act of 1986 empowered Agriculture Ministers to identify and administer Environmentally Sensitive Areas (ESAs). Where designations had previously drawn a distinction between the amenity, recreational, wildlife and historic value of features, the ESAs treated them as a unified whole, under the all encompassing concept of 'national environmental significance' (Smith, 1989).

Entirely voluntary in concept, ESAs offered payments from the Agricultural Budget where agreed management prescriptions were followed. The North Kent Marshes ESA was designated in the third round of 1993. Its coastal grazing marshes were of national and international importance for birds, as well as for its aquatic flora and archaeological interest. That environmental interest was threatened by such 'potentially damaging operations' as the lowering of water levels, levelling of land, increased fertiliser use, and conversion to arable and the consequent fragmentation of the traditional grazing-marsh system. As well as following the Codes of Good Agricultural Practice (as laid down by the Ministry), entry to grant aid was by way of two Tiers of Management. The first was intended to maintain the landscape and grasslands, with a further goal of enhancing the wildlife interest of the grassland, particularly for birdlife, by raising water levels in the ditches between 1 December and 30 April at not less than the mean field-level, so as to create extensive, shallow pools. An additional second tier was intended to encourage conversion of existing arable to permanent grass, so as to re-establish 'the characteristic wildlife found on the marshes'. Among other requirements, the sward had to contain species from an approved list within twelve months. It was not to be cut before 1 July of each year (Ministry of Agriculture, 1995).

The Agricultural Departments found themselves in a position analogous to that of the conservation bodies in the 1950s, as the number of designated areas grew. The policy instruments were in place, but it was far from clear whether there was the practical expertise and experience to ensure management prescriptions met their desired goals. In a report on *Agriculture and the Environment*, published in July 1984, the House of Lords' Select Committee on Science and Technology had called upon the NCC to strengthen considerably the research required for 'creative conservation' in lowland agricultural habitats (House of Lords, 1984). The Agriculture Departments emerged as by far the largest patrons of conservation research, their principal objective from the late 1980s being 'to develop, test and demonstrate effective, practical, least-cost farming practices that conserve and enhance landscape and biological diversity on farmland'. More specifically, studies commissioned to support the ESA programme have included the impact of water-level management on biodiversity and landscape features, and the drafting of guidelines on the restoration and maintenance of wet grasslands (MAFF, 1996).

— THE 'THEFT OF THE COUNTRYSIDE' —

In 1980, the same year as Marion Shoard's *The Theft of the Countryside* was published, there appeared a volume with a more prosaic title, *Britain's Future in Farming*. If the latter was written from a sharply differing perspective, its overriding preoccupation was the same, namely the need for reconciliation. If Shoard (1980) claimed farmers could no longer be trusted as 'custodians of the countryside', the latter volume believed it was a responsibility that had been foisted on them or, more particularly, on those farmers wrestling to make a living from the most marginal farmland. A main thesis of the medley of contributors under the senior editor, Sir Frank Engledow (the Emeritus Professor of Agriculture at Cambridge), was the need to raise agricultural production from a diminishing area of land. It was a matter of urgency, if not of national survival in an increasingly competitive and ultimately hungry world, that 'we must conserve the amount and improve the quality of our agricultural land'.

The threat to meeting that goal arose, in their judgement, from a lack of understanding on the part of the public that the primary use of land was for food production, and that the farmer was 'bound to try for the most efficient methods, if only to keep down costs and the consequent price of products' (Engledow and Amery, 1980). If such special pleading appeared borne of arrogance as to what had been achieved since the Second World War, it also gave voice to the inherent sense of insecurity felt by those associated with the industry. As a leading spokesmen, Sir Denys Bullard, remarked, in the annual parliamentary debate on agricultural subsidies in 1961,

> I have vivid memories of some of the Midlands counties as they were just before the war, where a great deal of arable land had decayed down to grass, where drainage had deteriorated ... the arts of arable husbandry had decayed completely. (PD, Commons, 640, 580–1.)

If such rhetorical warnings as to the consequences of any faltering of effort exaggerated the plight of land husbandry between the two world wars, the Chairman of the Countryside Commission, Sir Derek Barber, found the emotional strand of historical interpretation to be both powerful, and at times misleading, in portraying the inter-war countryside as unattractive. As Barber conceded, tumbled down fields and unkempt farmsteads might well have mocked farmers in their unfilled ambition and downright failure. And yet, compared with the 'scraped down' look of today's countryside, it was 'better clothed with woodland and trees', more diverse in character, and far better as a wildlife habitat. Nor was its 'shaggy' as opposed to 'manicured' appearance new. Compared with today, even the countryside of nineteenth-century High Farming would have appeared 'shaggy'. Despite the remarkable upheavals, say in property ownership, there had been little significant change in the appearance and character of the countryside until the 1940s. Even then, those most closely involved in what Barber called 'farming thought' and conservation did not fully appreciate the damage being wrought, say to wetlands, until the 1960s.

Perhaps a second major shift in perception occurred in the 1980s, as the conservation ethic was related to a wider political imperative to curb both agricultural-production costs and food surpluses. The conservation movement of the 1960s and 1970s, which had itself evolved from the pioneer efforts of the previous hundred years, found an 'economic' ally in the industry's own loss of confidence in the market place. Rather than an afterthought to agricultural considerations, environmental protection might offer both the rationale and the resources for reducing the intensity of land use (Barber, 1988).

But if the industry has argued that farmers, whatever their personal proclivities, were forced by business considerations to respond to the economic stimuli of demand for their produce during the post-war decades, a question arises. How far will they wish to take advantage of whatever opportunities are offered by the 'conservation tool', as an alternative to large-scale investment in intensive farming? Will such business opportunities be seized, or has the industry become so much a creature of its own sedulously promoted mythology that a stigma will continue to be attached to instances of 'shagginess', wherever and however they have come about? By their very nature, the traditional wetland areas will provide early evidence as to whether there is the will, let alone the competence.

References

ADAS Land Drainage Service, Ministry of Agriculture (n.d.). *History of Agricultural Drainage*, London.
Adkin, B. W. (1933). *Land Drainage in Britain*, Estates Gazette, London.
Allen, J. R. L. (1990). The Severn Estuary in South West Britain: its retreat under marine transgression and fine sediment regime, *Sedimentary Geology* **66**, 13-28.
Allen, J. R. L. (1996). The seabank on the Wentlooge Level, Gwent: date of set-back from documentary and pottery evidence, *Archaeology in the Severn Estuary* **7**, 67-84.
Allen, J. R. L. and Fulford, M. G. (1986). The Wentlooge Level: a Romano-British saltmarsh reclamation in south east Wales, *Britannia* **17**, 91-117.
Allen, J. R. L. and Fulford, M. G. (1990). Romano-British wetland reclamations at Longney, Gloucestershire, and evidence for the early settlement of the Inner Severn Estuary, *Antiquaries Journal* **80**, 288-326.
Allen, J. R. L. and Fulford, M. G. (1996). The distribution of south east Dorset black burnished category 1 pottery in south west Britain, *Britannia* **27**, 223-81.
Allen, R. C. (1992). *Enclosure and the Yeoman. The Agricultural Development of the South Midlands 1450-1850*, Clarendon, Oxford.
Allen, R. C. (1994). Agriculture during the Agricultural Revolution. In *The Economic History of Britain since 1700*, Vol. 1 (eds. Floud, R. and McClosky, D.), Cambridge University Press, Cambridge, 96-122.
Allison, K. J. (1976). *The East Riding of Yorkshire Landscape*, Hodder and Stoughton, London.
Andrews, G. H. (1853). *Modern Husbandry*, London.
Aston, M. (1988). *The Medieval Landscape of Somerset*, Somerset County Council, Taunton.
Atwood, G. (1964). A study of the Wiltshire Water Meadows, *Wiltshire Archaeological & Natural History Magazine* **58**, 403-13.
Aubrey, J. (1847). *Natural History of Wiltshire*, ed. J. Britton, London.
Avery, B. W., Findlay, D. C. and Mackney, D. (1975). *Soil Map of England*

and Wales, 11,000,000, Soils Survey, Southampton.

Bacon, R. N. (1844). *The Report on the Agriculture of Norfolk to which the prize was awarded by the Royal Agriculture Society of England*, London.

Bailey, J. (1810). *General View of the Agriculture of the County of Durham*, London.

Bailey, J. and Culley, G. (1805). *General View of the Agriculture of the County of Northumberland*, London.

Bailey, J. and Culley, G. (1813). *General View of the Agriculture of the Counties of Cumberland and Westmoreland*, London.

Baker, A. R. H. and Butlin, R. A. (1973). *Field Systems of the British Isles*, Cambridge University Press, Cambridge.

Baker, G. (1822). *History and Antiquities of the County of Northampton*, London.

Barber, D. (1988). The countryside: decline and renaissance, *Journal of the Royal Agricultural Society of England* **149**, 81–9.

Barnes, T. G. (1961). *Somerset 1625–1640*, Oxford University Press, Oxford.

Batchelor, T. (1813). *General View of the Agriculture of Bedfordshire*, London.

Beart, R. (1841). On the economical manufacture of draining-tiles and soles, *Journal of the Royal Agricultural Society of England* **2** (1841), 93–104.

Behre, K. E. and Jacomet, S. (1991). The ecological interpretation of archaeobotanical data. In *Progress in Old World Palaeoethnobotany* (eds Van Zeist, W., Wasylikowa, K. and Behre, K.), Balkema, Rotterdam, 81–108.

Bell, M. (1994). Field Survey and Excavation at Goldcliff, Gwent, 1994, *Archaeology in the Severn Estuary* **6**, 115–44.

Bell, M. (1995). Archaeology and nature conservation in the Severn Estuary, England and Wales. In *Wetlands: Archaeology and Nature Conservation* (eds Cox, M., Straker, V. and Taylor, D.), HMSO, London, 49–61.

Beresford, M. W. (1948). Ridge and furrow and the open fields, *Economic History Review* 2nd series **2**, 34–45.

Beresford, M. W. (1954). *The Lost Villages of England*, Lutterworth, London.

Beresford, M. W. and St. Joseph, J. K. (1979). *Medieval England from the Air* (2nd edn), Cambridge University Press, Cambridge.

Bettey, J. H. (1977). The Development of Water Meadows in Dorset during the Seventeenth Century, *Agricultural History Review* **25**, 37–43.

Billingsley, J. (1798). *General View of the Agriculture of Somerset*, London.

Bloom, A. H. V. (1944). *The Farm in the Fen*, Faber, London.

Bond, C. J. (1981). Otmoor. In Rowley, T. (ed.) 1981, 113–35.

Boswell, G. (1779). *A Treatise on Watering Meadows*, 2nd edn, London, 1790.

Bowden, P. J. (1985). Agricultural prices, wages, farm profits and rents. In Thirsk, J. (ed.) 1985b, 1–119.

Bowers, J. K. (1983). Cost-benefit analysis of wetland drainage. *Environment and Planning* **15**, 227–35.

Bowie, G. S. (1987). Watermeadows in Wessex: a re-evaluation for the period 1640–1850, *Agricultural History Review* **35**, 151–8.

Bowie, G. G. S. (1990). Northern wolds and Wessex downlands: contrasts in sheep husbandry and farming practices, 1770–1850, *Agricultural History Review* **38**, 117–26.

Bowles, N. (1988). *The Southern Wey: a guide*, Liphook.

Boys, J. (1813). *General View of the Agriculture of the County of Kent*, London.

BPP (British Parliamentary Papers), (1833), V, Select Committee on Agriculture.

BPP (British Parliamentary Papers), (1836), VIII, Select Committee appointed to enquire into the State of Agriculture.

BPP (British Parliamentary Papers), (1845), XVIII, Select Committee of the House of Lords... to enable Possessors of Settled Estates to charge such Estates... for the Purpose of Draining.

BPP (British Parliamentary Papers), (1873), XVI, Select Committee of the House of Lords on the Improvement of Land.

Bradley, R. (1727). *A General Treatise of Husbandry and Gardening*, London.

Bradley, R. J. (1975). Salt and settlement in the Hampshire Sussex borderland. In de Brisay, K. W. and Evans, K. A. (eds) 1975, 20–5.

Brady, N. C. (1974). *The Nature and Properties of Soils*, 8th edn, Macmillan, New York.

Brandon, P. (1971a). Agriculture and the effects of floods and weather at Barnhorne, Sussex, during the late Middle Ages, *Sussex Archaeological Collections* **109**, 69–93.

Brandon, P. (1971b). The origin of Newhaven and the drainage of the Lewes and Laughton Levels, *Sussex Archaeological Collections* **109**, 94–106.

Brandon, P. (1974). *The Sussex Landscape*, Hodder and Stoughton, London.

Brian, A. (1991). *The Lugg Meadows near Hereford*, Hereford Nature Trust, Hereford.

Brigham, T. (1990). The late Roman waterfront in London, *Britannia* **21**, 99–183.

Brooks, N. (1988). Romney Marsh in the early Middle Ages. In Eddison, J. and Green, C. (eds) 1988, 90–104.

Broomhead, R. A. (1991). *Archaeological Observations on the Lympsham Sewage Treatment Works Elimination Scheme*, unpublished Report: AC Archaeology.

Broughton, T. R. S. (1938). Roman Asia. In *An Economic Survey of Ancient Rome* (ed. Frank, T.), Vol 4, Johns Hopkins University Press, Baltimore, 499–916.

Brown, A. V. (1967). The last phase of the enclosure of Otmoor, *Oxoniensia* **32**, 43–52.

Buchanan, R. A. (1982). *Industrial Archaeology in Britain*, 2nd edn, Penguin, Harmondsworth.

Bullock, J. H. (1939). *The Norfolk Portion of the Chartulary of the Priory of St*

Pancras of Lewis, Norfolk Record Society, Norwich.

Burroughes, T. (1846). Farming in Cambridgeshire, *Journal of the Royal Agricultural Society of England* **7**, 54-72.

Burton, R. O. and J. M. Hodgson (1987). *Lowland Peat in England and Wales*, Soil Survey of England and Wales Special Survey no. 15, Harpenden.

Butlin, R. A. (1973). Field systems of Northumberland and Durham. In Baker, A. R. H. and Butlin, A. H. (eds) 1973, 93-144.

Caird, J. (1852). *English Agriculture in 1850-1*, London.

Caird, J. (1880). *The Landed Interest and the Supply of Food*, 4th edn, London.

Camden, W. (1586). *Britannica*, London.

Campbell, B. and Overton, M. (eds) (1991). *Land, Labour and Livestock: historical studies in European agricultural productivity*, Manchester University Press, Manchester.

Castle, D. A., McCunnall, J. and Tring, I. M. (1984). *Field Drainage, Principles and Practices*, Batsford, London.

Chambers, J. D. and Mingay, G. E. (1966). *The Agricultural Revolution 1750-1880*, Batsford, London.

Chase, M. (1992). Can history be green? A prognosis, *Rural history* **3** (2), 243-51.

Claridge, J. (1793). *General View of the Agriculture of Dorset*, London.

Cobbett, W. (1912). *Rural Rides*, Everyman Edn, London.

Coles, J. and Coles, B. (1996). *Enlarging the Past: The Contribution of Wetland Archaeology*, Society of Antiquaries of Scotland Monograph 11/Wetland Archaeological Research Project Occasional Paper 10.

Coles, B. P. L. and Funnel, B. M. (1981). *Holocene Palaeoenvironments of Broadland*, Special Publication of the International Association of Sedimentologists **5**, London.

Coles, J. M. and Orme, B. (1978). Structures south of Meare Island, *Somerset Levels Papers* **4**, 90-100.

Collingwood, R. G. and Wright, R. P. (1965). *The Roman Inscriptions of Britain*, Oxford University Press, Oxford.

Cook, H. F. (1993). Progress in water management in the lowlands, *Progress in Rural Policy and Planning* **3**, 91-103.

Cook, H. F. (1994). Field-scale water management in southern England to 1900, *Landscape History* **16**, 53-66.

Cook, H. F. and Dent, D. L. (1990). Modelling soil water supply to crops, *Catena* **17**, 25-39.

Cook, H. F. and Moorby, H. (1993). English marshlands reclaimed for grazing: a review of the physical environment, *Journal of Environmental Management* **38**, 55-72.

Copland, S. (1866). *Agriculture Ancient and Modern*, London.

Corbett, W. M. and Tatler, W. (1970). *Soils in Norfolk: Survey Record No. 1, Beccles North*, Harpenden.

Correl, D. L. and Weller, D. E., (1989). Factors limiting processes in freshwater wetlands: an agricultural primary stream riparian forest, In *Freshwater wetlands and wildlife* (eds Sharitz, R. R. and Gibbons, J. W.), USDOE symposium series no. 61.

Country Landowners Association, National Farmers Union and Royal Institution of Chartered Surveyors (1984). *Management Agreements in the Countryside*, London.

Cowan, M. (1982). *Floated Water Meadows in the Salisbury Area*, South Wiltshire Industrial Archaeology Society Historical Monograph **9**.

Cowell, R. W. and Innes, J. B. (1994). *The Wetlands of Merseyside*, Lancaster University Archaeological Unit, Lancaster.

Creyke, R. (1845). Some account of the process of warping, *Journal of the Royal Agricultural Society* **5**, 398-405.

Cunliffe, B. W. (1980). The evolution of Romney Marsh: a preliminary statement. In *Archaeology and Coastal Change* (ed. Thompson, F. H.), Society of Antiquaries, London, 37-55.

Cunliffe, B. (1988). Romney Marsh in the Roman period. In Eddison, J. and Green, C. (eds.) 1988, 83-7.

Curtis, L. F., F. M. Courtney and S. Trudgill (1976). *Soils of the British Isles*, Longman, London.

Darby, H. C. (1940). *The Medieval Fenland*, Cambridge University Press, Cambridge.

Darby, H. C. (1966). *The Draining of the Fens*, 2nd edn, Cambridge University Press, Cambridge.

Darby, H. C. (1983). *The Changing Fenland*, Cambridge University Press, Cambridge.

Davis, T. (1794). *General View of the Agriculture of the County of Wilts, with observations on the means of its improvement*, London.

de Brisay, K. W. and Evans, K. A. (1975). *Salt. The Study of an Ancient Industry*, Colchester Archaeological Group, Colchester.

de Gray Birch, W. (1881). *Memorials of St. Guthlac of Crowland*, London.

de Lavergne, L. (1855). *The Rural Economy of England, Scotland and Ireland*, Edinburgh and London.

Denison, J. (1840). On the Duke of Portland's water meadows at Clipstone Park, *Journal of the Royal Agricultural Society of England* **1**, 359-70.

Dent, D. L. (1986). *Acid Sulphate Soils: a baseline for research and development*, International Institute of Land Reclamation and Improvement, Wageningen.

Dent, D. L. (1992). Reclamation of acid sulphate soils, *Advances in Soil Science* **17**, 79-122.

Dent, D. L. and Scammell, R. P. (1981). Assessment of long-term irrigation need by integration of data for soil and crop characteristics and climate, *Soil Survey and Land Evaluation* **1**(3), 51-7.

Denton, J. B. (1842). General drainage and distribution of water, *Farmer's Magazine*, second series, **6**, 64.

Denton, J. B. (1855). *The Under-Drainage of Land: its progress and results*, London.

Denton, J. B. (1863). The effect of underdrainage on our rivers and arterial channels, *Journal of the Royal Agricultural Society of England* **24**, 573-89.

Denton, J. B. (1883). *Agricultural Drainage*, London.

Didsbury, P. (1988). Exploitation of the alluvium of the lower Hull valley in the Roman period. In *Humber Perspectives: a region through the ages* (eds Ellis, S. and Crowther, D. R.), Hull University Press, Hull, 199-210.

Dinnin, M. (1997). The drainage history of the Humberhead Levels. In Van de Noort, R. and Ellis, S. (eds), 19-30.

Doherty, J. (1985). Water over the meadows, *Natural World*, 1985, 17-19.

Dring, W. S. (1967). *The Fenland Story*, Cambridgeshire Libraries, Cambridge.

Drury, P. (1981). Medieval 'narrow rig' at Chelmsford and its possible implications, *Landscape History* **3**, 51-8.

Dugdale, W. (1662). *History of Inbanking and Drayning of Divers Fens and Marshes*, London.

Dulley, A. J. F. (1966). The Level and port of Pevensey in the Middle Ages, *Sussex Archaeological Collections* **104**, 26-45.

Dymond, D. (1990). *The Norfolk Landscape*, Hodder and Stoughton, London.

Eddison, J. (ed.) (1995). *Romney Marsh. The Debatable Ground*, Oxford University Committee for Archaeology Monograph 41, Oxford.

Eddison, J. and Green, C. (eds) (1988). *Romney Marsh. Evolution, Occupation, Reclamation*, Oxford University Committee for Archaeology Monograph 4, Oxford.

Electricity Council (1987). *Electricity Supply in the United Kingdom: a chronology*, 4th edn, Electricity Council, London.

Ellenberg, H. (1988). *Vegetation Ecology of Central Europe*, 4th edn, Cambridge University Press, Cambridge.

Elliott, G. (1973). Field systems of northwest England. In Baker, A. R. H. and Butlin, R. A (eds) 1973, 42-92.

Emanuelsson, U. and Moller, J. (1990). Flooding in Scania: a method to overcome the deficiency of nutrients in agriculture in the nineteenth century, *Agricultural History Review* **38**, 136.

Engledow, F., and Amery, L. (eds) (1980). *Britain's Future in Farming: principles of policy for British agriculture*, Geographical Publications, Berkhamsted.

Evans, G. Ewart (1953). *Ask the Fellows Who Cut the Hay*, Faber, London.

Evans, G. Ewart (1970). *Where Beards Wag All*, Faber, London.

Evans, H. (1845). Norfolk Draining, *Journal of the Royal Agricultural Society of England* **4**, 43-4.

Everitt, A. (1986). *Continuity and Colonisation. The Evolution of Kentish Settlement*, Leicester University Press, Leicester.

Eyre, S. R. (1955). The curving ploughland strip, *Agricultural History Review* **3**, 80-94.

FAO/ISRIC (1998). *World Reference Base for Soil Resources*, World Resources Report 84, Food and Agriculture Organisation/International Soil Reference and Information Centre, Rome.

Farrer, W. and Brownbill, J. (1907). *Victoria History of the County of Lancaster Vol III*, Institute of Historical Research, London.

Feltwell, J. (1992). *Meadows: a history and natural history*, Alan Sutton Publishing, Gloucester.

Findley, D. C., Colbourne, G. J. N., Cope, D. W., Harrod, T. R., Hogan, D. V. and Staines, S. J. (1984). *Soils and their uses in South West England*, Soil Survey of England and Wales Bulletin No. 14, Harpenden.

Finney, J. B., Finney, S. M. and James, N. (1995). Wind pumps in the Haddenham Level, *Proceedings of the Cambridge Antiquarian Society* **84**, 155-65.

Foard, G. R. and Hall, D. N. (1997). Note in English Heritage *Archaeology Review 1996-7*, London.

Fojt, W. J. (1995). The nature conservation importance of fens and bogs and the role of restoration. In *Restoration of Temperate Wetlands* (eds Wheeler, B. D., Shaw, S. C., Fojt W. J. and Robertson, R. A.), Wiley, London, 33-48.

Folkingham, W. (1610). *Feudographica*, London.

Fowler, P. J. (1983). *The Farming of Prehistoric Britain*, Cambridge University Press, Cambridge.

Frank, T. (1927). *An Economic History of Rome*, Johns Hopkins University Press, Baltimore.

French, C. (1996). Molluscan analysis. In Jackson, R. P. J. and Potter, T. W. (eds.) 1996, 639-54.

French, C. A. I. (1992). Fengate to Flag Fen: a summary of the soil and sediment analysis. *Antiquity* **66** p. 251.

French, H. F. (1879). *Farm Drainage*, New York.

Frere, S. (1987). *Britannia: A history of Roman Britain*, 3rd edn, Routledge and Kegan Paul, London.

Fulford, M. G. (1992). Iron Age to Roman: a period of radical change on the gravels. In Fulford, M. and Nichols, E. (eds) 1992, 22-38.

Fulford, M. G. (1996). *The Second Augustan Legion in the West of Britain. The Ninth Annual Caerleon Lecture*, Cardiff University, Cardiff.

Fulford, M. G., Allen, J. R. L. and Rippon, S. J. (1994). The settlement and drainage of the Wentlooge Level, Gwent: excavation and survey at Rumney Great Wharf 1992, *Britannia* **25**, 175-211.

Fulford, M. G. and Nichols, E. (eds) (1992). *Developing Landscapes of Lowland Britain. The Archaeology of the British Gravels: A Review*, Society of Antiquaries Occasional Paper **14**.

Gardiner, J. S. (1932). *The Natural History of Wicken Fen*, Bowes and Bowes, Cambridge.

Gardiner, M. (1988). Medieval settlement and society in the Broomhill area and excavations at Broomhill church. In Eddison, J. and Green, C. (eds) 1988, 112-27.

Gardiner, M. (1995). Medieval farming and flooding in the Brede Valley. In Eddison, J. (ed) 1995, 127-37.

George, M. (1992). *The Land Use, Ecology and Conservation of Broadland*, Packard, Chichester.

Gerard, T. (1980). *General Description of Dorset*, ed. R. Legg, London.

Glyde, J. (1856). *Suffolk in the Nineteenth Century*, London.

Godwin, H. (1978). *Fenland: its ancient past and uncertain future*, Cambridge University Press, Cambridge.

Gooch, W. (1813). *General View of the Agriculture of the County of Lincolnshire*, London.

Goodland, N. L. (1958). New ways with old meadows, *Country Life* 9 Oct., 762-4.

Gray, H. L. (1915). *English Field Systems*, Cambridge, Mass.

Green, F. H. W. (1979a). *Field Drainage in Europe: a quantitative survey*, Institute of Hydrology Report no. 57, London.

Green, F. H. W. (1979b). Field under-drainage and the hydrological cycle. In *Man's Impact on the Hydrological Cycle in the United Kingdom* (ed. G. E. Hollis) Geo Abstracts, Norwich, 9-18.

Green, R. D. (1968). *Soils of Romney Marsh*, Soil Survey of England and Wales, Harpenden.

Grigg, D. B. (1965). An index of regional change in English farming, *Transactions of the Institute of British Geographers* **36**, 55-67.

Gross, A. and Butcher, A. (1995). Adaptation and investment in the age of the great storms: agricultural policy on the manors of the principal lords of the Romney Marshes and the marshland fringe c.1250-1320. In Eddison, J. (ed.) 1995, 107-17.

Hall, D. (1972). Modern surveys of medieval field systems, *Bedfordshire Archaeological Journal* **7**, 53-66.

Hall, D. (1981). The changing landscape of the Cambridgeshire silt fens, *Landscape History* **3**, 37-49.

Hall, D. (1981). The origins of open-field agriculture – the archaeological fieldwork evidence. In Rowley, T. (ed.) 1981, 22-8.

Hall, D. (1982). *Medieval Fields*, Shire Publications, Aylesbury.

Hall, D. (1987). *The Fenland Project, Number 2: Cambridgeshire Survey, Peterborough to March, East Anglian Archaeology* 35.

Hall, D. (1992). *The Fenland Project, Number 6: The South-Western Cambridgeshire Fenlands, East Anglian Archaeology* 56.

Hall, D. and Coles, J. (1994). *Fenland Survey. An Essay in Landscape and*

Persistence, English Heritage, London.
Hall, D. (1995). *The Open Fields of Northamptonshire*, Northamptonshire Record Society, Vol. 38.
Hall, D. (1996). *The Fenland Project, Number 10: Cambridgeshire Survey, Isle of Ely and Wisbech, East Anglian Archaeology* **79**.
Hall, D., Wells, C. E. and Huckerby, E. (1995). *The Wetlands of Greater Manchester*, Lancaster University Archaeological Unit, Lancaster.
Hallam, H. E. (1965). *Settlement and Society. A Study of the Early Agrarian History of South Lincolnshire*, Cambridge University Press, Cambridge.
Hallam, H. E. (1981). *Rural England 1066-1348*, Fontana, London.
Hallam, H. E. (1988). Drainage techniques. In *The Agrarian History of England and Wales, Volume 2 (1042-1350)*, (ed. Hallam, H. E.), Cambridge University Press, Cambridge.
Hallam, S. J. (1970). Settlement round the Wash. In Phillips, C. W. (ed.) 1970, 22-113.
Hardy, T. (1968). *Return of the Native*, Macmillan edn, London.
Harrison, M. J., Mead, W. R. and Pannett, D. J. (1965). A Midland ridge-and-furrow map, *Geographical Journal* **131**, 366-9.
Harvey, N. (1980). *The Industrial Archaeology of farming in England and Wales*, Batsford, London.
Harvey, M. (1981). The origin of planned field systems in Holderness, Yorkshire. In Rowley (ed.) 1981, 184-201.
Harvey, M. (1982). Regular open-field systems on the Yorkshire Wolds, *Landscape History* **4**, 28-39.
Havinden, M. (1981). *The Somerset Landscape*, Hodder and Stoughton, London.
Hayes, P. P. and Lane, T. W. (1992). *The Fenland Project, Number 5: Lincolnshire Survey, The South-West Fens, East Anglian Archaeology* **55**.
Hazelden, J. (1990). *Soils in Norfolk: Survey Record No. 115*, Silsoe, Harpenden.
Heathcote, J. M. (1877). *Scoopwheel and Centrifugal Pump*, Cambridge.
Hey, D. (1984). Yorkshire and Lancashire. In Thirsk, J. (ed.) 1985a, 59-88.
Hill, M. O. (1993). *TABLEFIT Version 0.0 for Identification of Vegetation Types*, Institute of Terrestrial Ecology, Huntingdon.
Hillel, D. (1982). *Introduction to Soil Physics*, Academic Press, London.
Hills, R. L. (1967). *Machines, Mills and Uncountable Costly Necessities*, Goose and Son, Norwich.
Hodge, C. A. H., Burton, R. G. O., Corbett, W. M., Evans, W. M. and Seale, R. S. (1984). *Soils and their use in Eastern England*, Soil Survey of England and Wales Bulletin no. 13, Harpenden.
Holderness, B. A. (1971). Capital formation in agriculture. In *Aspects of Capital Formation in Great Britain 1750-1850* (eds Higgins, J. P. P. and Pollard, S.), London, 159-83.

Holderness, B. A. (1972). Landlord's capital formation in East Anglia 1750-1870, *Economic History Review, Second Series* **25**, 434-47.
House of Commons (1985). *First Report from the Environment Committee. Operation and Effectiveness of Part II of the Wildlife and Countryside Act*, Volumes 6 and 11, London.
House of Lords (1984). *Fourth Report from the Select Committee on Science and Technology. Agricultural and Environmental Research*, Volumes 271-2, London.
Hume, L. (1993). *Archaeological Evaluation of Land at Rust Bridge, Kenn, Avon*, unpublished report, Avon Archaeological Unit.
Hutchinson, J. N. (1980). The record of peat wastage in the East Anglian Fenlands at Holme 1848-1978 AD, *Journal of Ecology* **68** (1), 229-49.
Institute of Geological Sciences/Southern Water Authority (1979). *Hydrological Map of Hampshire and the Isle of Wight*, National Environmental Research Council, London.
Jackson, R. P. J. and Potter, T. W. (1996). *Excavations at Stonea, Cambridgeshire 1980-85*, British Museum Press, London.
James, J. F. and Bettey, J. H. (1993). *Farming in Dorset: The Diary of James Warne 1758 and the Letters of George Boswell 1787-1805*, Dorset Record Society, Dorchester.
Johnson, E. A. G. (1954). Land drainage in England and Wales, *Proceedings of the Institution of Civil Engineers* **3**(3), 601-51.
Johnstone, J. (1801). *An Account of the Mode of Draining Land According to the System Practised by Mr. Joseph Elkington*, 2nd edn, London.
Joint Select Committee (1927). *Report on the Ouse Drainage Bill*, Parliamentary Papers VI, London.
Jones, B. and Mattingly, D. (1990). *An Atlas of Roman Britain*, Basil Blackwell, Oxford.
Jones, E. L. (1960). Eighteenth-century changes in Hampshire chalkland farming, *Agricultural History Review*, **8**, 5-19.
Kain, R. J. B. and Mead, W. R. (1976). Ridge-and-furrow in Kent, *Archaeologia Cantiana* **92**, 165-71.
Kain, R. J. P. and Mead, W. R. (1977). Ridge-and-furrow in Cambridgeshire, *Proceedings of the Cambridgeshire Antiquarian Society* **67**, 131-7.
Kain, R. J. P. (1986). *An Atlas and Index of the Tithe Files of Mid-Nineteenth-Century England and Wales*, Cambridge University Press, Cambridge.
Keil, I. J. E. (1964). *The Estates of Glastonbury Abbey in the Later Middle Ages*, unpublished PhD thesis, University of Bristol.
Kent, N. (1793). *Hints to Gentlemen of Landed Property*, London.
Kent, N. (1794). *General View of the Agriculture of the County of Norfolk*, London.
Kerridge, E. (1951). Ridge and furrow and agrarian history, *Economic History Review* 2nd series **4**, 14-36.

Kerridge, E. (1953a). The sheep fold in Wiltshire and the floating of the watermeadows, *Economic History Review* **6**, 282-9.
Kerridge, E. (1953b). *Surveys of the Manors of Philip, 1st Earl of Pembroke*, Wiltshire Archaeological and Natural History Society, Devizes.
Kerridge, E. (1953c). The floating of the Wiltshire water meadows, *Wiltshire Archaeology and Natural History Magazine* **15**, 105-18.
Kerridge, E. (1955). A reconsideration of some former husbandry practices, *Agricultural History Reviews* **3**, 32-40.
Kerridge, E. (1967). *The Agricultural Revolution*, George Allen and Unwin, London.
Kerridge, E. (1973). *The Farmers of Old England*, George Allen and Unwin, London.
Lambrick, G. (1992). Iron Age to Roman: a period of radical change on the gravels. In Fulford, M. And Nichols, E. (eds.) 1992, 78-105.
Lane, T. W. (1993). *The Fenland Project, Number 8: Lincolnshire Survey, The Northern Fen-edge*, East Anglian Archaeology 66.
Leah, M. and Crowson, A. (1993). Norfolk Archaeological Unit: the Fenland Management Project, *Fenland Research* **8**, 43-50.
Leah, M., Wells, C. E., Appleby, C. and Huckerby, E. (1997). *The Wetlands of Cheshire*, Lancaster University Archaeological Unit, Lancaster.
Lefevre, C. S. (1836). *Remarks on the Present State of Agriculture*, London.
Liddiard, R. (1997). *Ridge and Furrow in Norfolk: ploughing and subsequent land use*, unpublished MA dissertation, Centre for East Anglian Studies, University of East Anglia.
Lobel, M. (ed.) (1957). The Victoria History of the County of Oxfordshire, Vol. V, London.
Loch, J. (1820). *An Account of the Improvements on the Estates of the Marquis of Stafford*, London.
Locock, M. (1996). Hill farm, Goldcliff: a field evaluation of the proposed Gwent Levels Nature Reserve, *Archaeology in the Severn Estuary* **1966**, 59-66.
Lowe, R. (1798). *General View of the Agriculture of the County of Nottingham*, London.
Mackney, D. (1975). Soil Maps and Classification. In Soils and Field Drainage (ed. Thomasson, A. J.), Soil Survey Technical Monograph no. 7, Harpenden, pp 35-48.
Maitland, (1897). *Domesday Book and Beyond*, London.
Malcolmson, R. W. (1981). *Life and Labour in England 1700-1780*, Hutchinson, London.
Marshall, E. J. P. (1984). The ecology of a land drainage channel II: biology, chemistry and submerged weed control, *Water Research* **18**, vii, 817-25.
Marshall, T. J. (1959). *Relations between Water and Soil*, Technical Communication no. 50, Commonwealth Agricultural Bureaux, Harpenden, England.

Marshall, W. (1796). *The Rural Economy of the Midland Counties*, Volume I, London.
Mavor, W. (1809). *General View of the Agriculture of Berkshire*, London.
McDonald, R. (1908). *Agricultural Writers from Sir Waler of Henley to Arthur Young 1200-1800*, H. Cox, London.
Mead, W.R. (1954). Ridge-and-furrow in Buckinghamshire, *Geographical Journal* **120**, 34-42.
Mead, W. R. and Kain, R. J. P. (1976-7). Ridge-and-furrow in Kent, *Archaeologia Cantiana* **92**, 165-71.
Middleton, J. (1813). *General View of the Agriculture of the County of Middlesex*, London.
Middleton, R., Wells, C. and Huckerby, E. (1995). *The Wetlands of North Lancashire*, Lancaster University Archaeological Unit, Lancaster.
Miller, E. (1988). Northern England. In *Agrarian History of England and Wales, Volume 2 (1042-1350)*, (ed. Hallam, H. E.), Cambridge University Press, Cambridge.
Miller, S. H. and Skertchly, S. B. J. (1878). *The Fenland; past and present*, Longmans, London.
Millett, M. (1990). *The Romanization of Britain*, Cambridge University Press, Cambridge.
Mingay, G. E. (1994). *Land and Society in England 1750-1980*, Longman, London.
Ministry of Agriculture, Fisheries and Food (1926). *Report of the Commission Appointed in Connection with the Ouse Drainage District*, Parliamentary Papers, 1926, XV, Cmd 2572, London.
Ministry of Agriculture, Fisheries and Food (1966). *Grass and Grassland*, Bulletin number 154, HMSO, London.
Ministry of Agriculture, Fisheries and Food (1995). *Environmentally Sensitive Areas. The North Kent Marshes. Guidelines for Farmers*, London.
Ministry of Agriculture, Fisheries and Food (1996). *Research Strategy 1996-2000*, HMSO, London.
Mitchell, G. S. (1894). *A Handbook of Land Drainage*, Land Agents Record, London.
Moon, H. P., and Green, F. H. W. (1940). Watermeadows in Southern England. In *The Land of Britain, The Report of the Land Utilisation Survey of Britain* (ed. L. D. Stamp), Pt. 89, Appendix 2, 373-90.
Moore, H. I. and Burr, S. (1948). The control of rushes on newly re-seeded land in Yorkshire, *Journal of the British Grasslands Society* **3**, 283-90.
Moore, I. (1966). *Grass and Grasslands*, Collins, London.
Morris, J. (1989). *Land Drainage: agricultural benefits and impacts*. Technical papers from the annual Symposium of Institute of Water and Environmental Management, York.
Mountford, O. (1997). *The Wet Grassland Guide*, Royal Society for the

Protection of Birds, Sandy, Bedfordshire.

Murphy, P. (1993a). Environmental archaeology: second progress report, *Fenland Research* **8**, 35-9.

Murphy, P. (1993b). Anglo-Saxon arable farming on the silt fens – preliminary results, *Fenland Research* **8**, 75-9.

Murphy, P. (1994). Environmental archaeology: third progress report. *Fenland Research* **9**, 26-9.

Murray, A. (1813). *A General View of the County of Warwick*, London.

Murray, K. A. H. (1955). *Agriculture*, HMSO, London.

National Rivers Authority (1994). *Water Resources in Anglia*, National Rivers Authority, Peterborough.

Nature Conservancy Council (1977). *Third Report*, London.

Nature Conservancy Council (1985). *Nature Conservation in Great Britain*, Nature Conservancy Council, Peterborough.

Newbould, C., Honnor, J. and Buckley, K. (1989). *Nature Conservation and the Management of Drainage Channels*, Nature Conservancy Council and the Association of Drainage Authorities, Peterborough.

Newson, M. (1992). *Land, Water and Development*, Routledge, London.

Orwin, C. S. (1929). *The Reclamation of Exmoor Forest*, Oxford University Press, London.

Orwin, C. S. and Orwin, C. S. (1938). *The Open Fields*, Clarendon Press, Oxford.

Oschinsky, D. (1971). *Walter of Henley and Other Treatises on Estate Management and Accounting*, Clarendon Press, Oxford.

Overton, M. (1996). *Agricultural Revolution in England: the transformation of the agrarian economy 1500-1850*, Cambridge University Press, Cambridge.

Owen, A. E. B. (1981). *The Records of a Commission of Sewers for Wiggenhall 1319-1324*, Norfolk Record Society, Norwich.

Palmer, R. (1996). Air photo interpretation and the Lincolnshire Fenland, *Landscape History* **18**, 5-16.

Parkes, J. (1843). Report on drain-tiles and drainage, *Journal of the Royal Agricultural Society of England* **4**, 369-79.

Parkes, J. (1845a). On reducing the cost of permanent drainage, *Journal of the Royal Agricultural Society of England* **6**, 125-9.

Parkes, J. (1845b). Report on the exhibition of implements at the Shrewsbury meeting in 1845, *Journal of the Royal Agricultural Society of England* **6**, 303-23.

Parkes, J. (1846). On draining, *Journal of the Royal Agricultural Society of England* **7**, 249-72.

Parkinson, R. (1813). *General View of the Agriculture of the County of Huntingdon*, London.

Parkinson, M. (1980). Salt marshes of the Exe Estuary, *Report and Transactions of the Devon Association for the Advancement of Science* **112**, 17-41.

Parkinson, R. J. (1995). Soil management. In *The Agricultural Notebook* (ed. Soffe, R. J.), 19th edn, Blackwell Science, London.

Phillips, A. D. M. (1975). Underdraining and agricultural investment in the Midlands in the mid-nineteenth century. In *Environment, Man and Economic Change* (eds. Phillips, A. D. M. and Turton, B. J.), Longman, London, 253-74.

Phillips, A. D. M. (1989). *The Underdraining of Farmland in England during the Nineteenth Century*, Cambridge University Press, Cambridge.

Phillips, A. D. M. (1996). *The Staffordshire Reports of Andrew Thompson to the Inclosure Commissioners, 1858-68*, Staffordshire Record Society, Stafford.

Phillips, C. W. (ed.) (1970). *The Fenland in Roman Times*, Royal Geographical Society Research Series 5, London.

Pitt, W. (1813). *General View of the Agriculture of the County of Leicestershire*, London.

Plater, A. and Long, A. (1995). The morphology and evolution of Denge Beach and Denge Marsh. In Eddison, J. (ed.) 1995, 8-36.

Plot, R. (1705). *The Natural History of Oxfordshire* (1st edn 1676), London.

Postgate, M. R. (1973). Field systems of East Anglia. In Baker, A. R. H. and Butlin, R. A. (eds) 1973, 281-324.

Potter, C., Cook, H. F. and Norman, C. (1993). The targeting of rural environmental policies: an assessment of agri-environmental schemes in the UK, *Journal of Environmental Planning and Management* **36(2)**, 199-215.

Potter, T. W. (1981). The Roman occupation of the central Fenland, *Britannia* **12**, 79-133.

Potter, T. W. (1989). The Roman Fenland: a review of recent work. In *Research on Roman Britain 1969-89* (ed. Todd, M.), *Britannia* Monograph Series **11**, 147-73.

Potter, T. W. and Jackson, R. P. J. (1982). The Roman site of Stonea, Cambs, *Antiquity* **217**, 111-20.

Priest, Rev. St J. (1813). *General View of the Agriculture of Buckinghamshire*, London.

Public Local Inquiry (1978). *Proposals by the Southern Water Authority for Drainage of Amberley Wild Brooks*, unpublished Inspector's Report and Evidence, London.

Pugh, R. B. (ed.) (1968). The Victoria History of the County of Wiltshire, Vol. IX, London.

Purseglove, J. (1988). *Taming the Flood*, Oxford University Press, Oxford.

Pusey, P. (1841). Editorial note, *Journal of the Royal Agricultural Society of England* **2**, 103.

Pusey, P. (1842). On the progress of agricultural knowledge during the last four years, *Journal of the Royal Agricultural Society of England* **3**, 169-217.

Pusey, P. (1843). Evidence of the antiquity, cheapness and efficacy of thorough-draining or land-ditching as practised throughout the counties of

Suffolk, Hertford, Essex and Norfolk, *Journal of the Royal Agricultural Society of England*, **4**, 23-49.

Pusey, P. (1845). Remarks on the foregoing evidence [i.e. re Evans (1845)], *Journal of the Royal Agricultural Society of England* **4**, 44.

Pusey, P. (1846). Editor's footnote to Roal, 1846.

Pusey, P. (1849). On the theory and practice of water meadows, *Journal of the Royal Agricultural Society of England* **10**, 462-79.

Raadsma, S. (1974). Current drainage practices in flat areas of humid regions of Europe. In *Drainage for Agriculture* (ed. Van Schilfgaarde, J.), American Society of Agronomy, Madison, Wisconsin, 115-39.

Rackham, O. (1986). *The History of the Countryside*, Dent, London.

Ratcliffe, D. A. (1977). *Nature Conservation Review*, Vols I and II, Cambridge University Press, Cambridge.

Raynbird, W. and H. (1849). *On the Farming of Suffolk*, London.

Read, C. S. (1858). Recent improvements in Norfolk Farming, *Journal of the Royal Agricultural Society of England* **19**, 265-311.

Redclift, M. (1994). Reflections on the 'sustainable development debate', *International Journal of Sustainable Development and World Ecology* **1**, 3-21.

Reeve, R. C. and N. R. Fausey (1974). Draining and timeliness of farming operations. In *Drainage for Agriculture* (ed. Van Schilfgaarde, J.), American Society of Agronomy, Madison, Wisconsin, 55-66.

Reeves, A. (1995). Romney Marsh: The fieldwalking evidence. In Eddison, J. (ed.) 1995, 78-98.

Richardson, J. J., Jordan, A. G. and Kimber, R. H. (1978). Lobbying, administrative reform and policy styles: the case of land drainage, *Political Studies* **26**, 47-64.

Richardson, S. J. and Smith, J. (1977). Peat wastage in the East Anglian Fens, *Journal of Soil Science* **28**, 485-9.

Richmond, I. A. (1963). *Roman Britain*, 2nd edn, Penguin, Harmondsworth.

Rippon, S. (1994a). The Roman settlement and landscape at Kenn Moor, North Somerset: report on survey and excavation 1993/4, *Archaeology in the Severn Estuary 1994*, 21-34.

Rippon, S. (1994b). Medieval wetland reclamation. In *The Medieval Landscape of Wessex* (eds. Aston, M. and Lewis, C.), Oxbow Books, Oxford, 239-53.

Rippon, S. (1995a). Human-environment relations in the Gwent Levels: ecology and the historic landscape in a coastal wetland. In *Wetlands: Archaeology and Nature Conservation* (eds Cox, M., Straker, V. and Taylor, D.), HMSO, London, 62-74.

Rippon, S. (1995b). The Roman settlement and landscape at Kenn Moor, North Somerset: interim report on survey and excavation 1994/5, *Archaeology in the Severn Estuary 1995*, 35-47.

Rippon, S. (1996a). *The Gwent Levels: the evolution of a wetland landscape*, Council for British Archaeology Research Report 105, CBA, London.

References

Rippon, S. (1996b). Roman and medieval settlement on the North Somerset Levels: survey and excavation at Banwell and Puxton, 1996, *Archaeology in the Severn Estuary 1996*, 67-84.

Rippon, S. (1997a). *The Severn Estuary: landscape evolution and wetland reclamation*, Leicester University Press, Leicester.

Rippon, S. (1997b). Roman settlement and salt production on the Somerset coast: the work of Samuel Nash - a Somerset archaeologist and historian 1913-1985, *Somerset Archaeology and Natural History* **140**.

Rippon, S. (Forthcoming a). *Transformation and Control: the exploitation and modification of coastal wetlands in Britain and North West Europe*.

Rippon, S. (Forthcoming b). The North Somerset Levels: survey and excavation in a Romano-British coastal wetland, *Britannia*.

Roal, J. (1846). On the converting of mossy hillside to catch meadows, *Journal of the Royal Agricultural Society of England* **6**, 518-22.

Roberts, B. K. (1973). Field systems of the West Midlands. In Baker, A. R. H. and Butlin, R. A. (eds) 1973, 195-205.

Roberts, B. K. and Wrathmell, S. (1995). *Terrain and Local Settlement Mapping*, English Heritage, London.

Robinson, G. (1988). Sea defence and land drainage on Romney Marsh. In Eddison, J. and Green, C. (eds) 1988, 162-6.

Robinson, M. (1992). Environmental archaeology of the river gravels: past achievements and future directions. In Fulford, M. and Nichols, E. (eds.) 1992, 22-38.

Roden, D. (1973). Field systems of the Chiltern Hills and their environs. In Baker, A. R. H. and Butlin, R. A. (eds) 1973, 325-76.

Rodwell, J. S. (ed.) (1991). *British Plant Communities. Volume 2. Mires and Heath*, Cambridge University Press, Cambridge.

Rodwell, J. S. (ed.) (1992). *British Plant Communities. Volume 3. Grasslands and Montane Communities*, Cambridge University Press, Cambridge.

Rodwell, J. S. (ed.) (1995). *British Plant Communities. Volume 4. Aquatic Communities, Swamps and Tall-herb Fens*, Cambridge University Press, Cambridge.

Rose, S. C. and Armstrong, A. C. (1992). Agricultural field drainage and the changing environment, *SEESOIL* **8**, 53-68.

Rosen, A. (1978). Babraham. In *The Victoria County History of Cambridgeshire* **6**, Institute of Historical Research, London, 19-28.

Roseveare, J. C. A. (1932). Land drainage in England and Wales, *Transactions of the Institution of Water Engineers* **37**, 178-207.

Rostovtzeff, M. (1957). *The Social and Economic History of the Roman Empire*, 2nd edn, Oxford University Press, Oxford.

Rowley, T. (ed.) (1978). *The Evolution of Marshland Landscapes*, Oxford University Depatment for Extramural Studies, Oxford.

Rowley, T. (ed.) (1981). *The Origins of Open-Field Agriculture*, Croom Helm, London.

Royal Commission on Land Drainage (1927). *Report*, Parliamentary Papers, 1927, X, Cmd 2993, London.

Royal Society for Nature Conservation, (1991). *Losing Ground. Vanishing meadows, the case for extending the ESA principle*, The Wildlife Trust Partnership, Lincoln.

Royal Society for the Protection of Birds (1992). *Time for a Greater Thames*, RSPB, Sandy, Bedfordshire.

Royal Society MSS, Classified Papers, 1660-1740, 10/3/10, Report to the Georgical Committee.

Rudge, T. (1807). *General View of the Agriculture of Gloucestershire*, London.

Ruegg, L. H. (1854). Farming in Dorset, *Journal of the Royal Agricultural Society of England* **15**, 414-20.

Russell, E. W. (1973). *Soil Conditions and Plant Growth*, 10th edn, Longman, Harlow.

Russett, V. (1989). A Romano-British site at Rooksbridge, Somerset, *Axbridge Archaeological and Local History Society Journal*, 37-49.

Salmon, N. (1728). *The History of Hertfordshire*, London.

Salway, P. (1970). The Roman Fenland. In Phillips, C. W. (ed.) 1970, 1-21.

Salway, P. (1981). *Roman Britain*, Oxford University Press, Oxford.

Salzmann, L. F. (1910). The inning of Pevensey Levels, *Sussex Archaeological Collections* **53**, 32-60.

Scarth, H. (1885). The Roman Villa at Wemberham in Yatton, *Somerset Archaeol. Nat. Hist.* **31**, 1-9.

Scotter, C. N. G., Wade, P. M., Marshall, E. J. P. and Edwards, R. W. (1977). The Monmouthshire Levels' drainage system: its ecology and relation to agriculture, *Journal of Environmental Management* **5**, 75-86.

Seale, R. S. (1975). *Soils of the Chatteris District of Cambridgeshire*, Soil survey of England and Wales Special Survey 9, Harpenden.

Secretaries of State for Northern Ireland, Scotland and Wales, and Minister of Agriculture (1975). *Food from our own Resources*, HMSO, London.

Seebohm, F. (1883). *The English Village Community*, London.

Shannan, I. (1989). Holocene coastal movements and sea level changes in Britain, *Journal of Quarterly Science* **4(1)**, 77-89.

Shaw, E. M. (1983). *Hydrology in Practice*, Chapman & Hall, London.

Sheail, J. (1971). The formation and maintenance of water-meadows in Hampshire, England, *Biological Conservation*, 3(2), 101-6.

Sheail, J. (1993). Green history – the evolving agenda, *Rural History* **4(2)**, 209-23.

Sheail, J. (1995). Nature protection, ecologists and the farming context: a UK perspective, *Journal of Rural Studies* **11**, 79-88.

Sheail, J. (1996). Town wastes, agricultural sustainability and Victorian sewage, *Urban History* **23(2)**, 189-210.

Shennan, I. (1989). Holocene crustal movements and sea-level changes in Great Britain, *Journal of Quaternary Science* **4**, 77-89.

Sheppard, J. A. (1958). *The Draining of the Hull Valley*, East Yorks Local History Society, York.
Sheppard, J. A. (1966). *The Draining of the Marshlands of South Holderness and the Vale of York*, East Yorks Local History Society, York.
Sheppard, J. A. (1973). Field systems of Yorkshire. In Baker, A. R. H. and Butlin, R. A. (eds) 1973, 145-87.
Shiel, R. S. (1991). Soil fertility in the pre-fertiliser era. In *Land, Labour and Livestock* (eds. Campbell, B. and Overton, M.), Manchester University Press, Manchester, 51-77.
Shoard, M. (1980). *The Theft of the Countryside*, Maurice Temple Smith, London.
Short, B. M. (1985). South East England: Kent, Surrey and Sussex. In Thirsk, J. (ed.) 1985a, 270-313.
Silvester, R. J. (1988). *The Fenland Project, Number 3: Norfolk Survey, Marshland and the Nar Valley*, East Anglian Archaeology 45.
Silvester, R. J. (1989). Ridge and furrow in Norfolk, *Norfolk Archaeology* **40**, 286-96.
Silvester, R. J. (1991). *The Fenland Project, Number 4: Norfolk Survey, The Wissey Embankment and Fen Causeway*. East Anglian Archaeology 35.
Silvester, R. J. (1993). 'The addition of more-or-less undifferentiated dots to a distribution map'? The Fenland Project in retrospect. In *Flatlands and Wetlands: Current Themes in East Anglian Archaeology*. East Anglian Archaeology 58 (ed. Gardiner, J.), 24-39.
Simmons, B. B. (1979). The Lincolnshire Car Dyke: navigation or drainage?, *Britannia* **10**, 183-96.
Simmons, B. B. (1980). Iron Age and Roman coasts around the Wash. In *Archaeology and Coastal Change* (ed. Thompson, F. H.), Society of Antiquaries Occasional Papers, New Series 1, London, 56-73.
Smith, E. A. G. (1989). Environmentally sensitive areas: a successful UK initiative, *Journal of the Royal Agricultural Society of England* **150**, 30-43.
Smith, J. (1831). *Remarks on Thorough Draining and Deep Ploughing*, Stirling.
Smith, L. P. (1976). *The Agricultural Climate of England and Wales*, HMSO, London.
Smith, L. P. and Trafford, B. D. (1976). *Climate and Drainage*, MAFF, HMSO, London.
Smith, R. A. L. (1940). Marsh embankment and sea defences in medieval Kent, *Economic History Review* **10**, 29-37.
Smith, R. (1851). Some account of the formation of hill-side catch-meadows on Exmoor, *Journal of the Royal Agricultural Society of England* **12**, 139-48.
Smith, W. (1806). *Observations on the Utility, Form and Management of Watermeadows*, Norwich.
Soil Survey of England and Wales. (1983). *Soil Map of England and Wales, 1:250,000*, Soil Survey, Southampton.
Soil Survey of England and Wales (1984). *Regional Bulletins 10, Northern*

England; 11, *Wales;* 12, *Midland and West of England;* 13, *East of England;* 14, *South West England;* 15, *South East England,* Soil Survey, Harpenden.

Spring, D. (1963). *The English Landed Estate in the Nineteenth Century: its administration,* Johns Hopkins University Press, Baltimore.

Spurrell, F. C. J. (1885). Early sites and embankments on the margins of the Thames Estuary, *Archaeological Journal* **42**, 269-302.

Stace, C. (1992). *New Flora of the British Isles,* Cambridge University Press, Cambridge.

Stallibrass, S. (1996). Animal bones. In Jackson, R. P. J. and Potter, T. W. (eds.) 1996, 587-612.

Stephens, H. (1851). *The Book of the Farm,* Edinburgh.

Stevenson, W. (1812). *General View of the Agriculture of Dorset,* London.

Stevenson, W. (1815). *General View of the County of Lancashire,* London.

Stewart, J. G. (1919). *Report to the Board of Agriculture on the Drainage of the Valleys of the Parrett, Tone, Cary, Brue and Axe,* Bridgewater.

Tansley, A. G. (1949). *The British Islands and their Vegetation, Volume 2,* Cambridge University Press, Cambridge.

Tatton-Brown, T. (1988). The topography of the Walland Marsh area between the eleventh and thirteenth centuries. In Eddison, J. and Green, C. (eds.) 1988, 105-11.

Taylor, C. C. (1970). *Dorset,* Hodder and Stoughton, London.

Taylor, C. C. (1973). *The Cambridgeshire Landscape,* Hodder and Stoughton, London.

Taylor, C. C. (1975). *Fields in the English Landscape,* Dent, London.

Taylor, C. C. (1981). The Drainage of Burwell Fen, Cambridgeshire, 1840-1950, In Rowley, T. (ed.) 1978, 158-77.

Taylor, C. (1983). *Village and Farmstead: a history of rural settlement in England,* George Philip, London.

Thirsk, J. (1953a). The Isle of Axholme before Vermuyden, *Agricultural History Review* **1**, 16-28.

Thirsk, J. (1953b). *Fenland Farming in the Sixteenth Century,* Leicester University Department of English Local History Occasional Paper 3, Leicester.

Thirsk, J. (ed.) (1985a). *The Agrarian History of England and Wales Volume V.1, Regional Farming Systems,* Cambridge University Press, Cambridge.

Thirsk, J, (ed.) (1985b) *The Agrarian History of England and Wales Volume V.2, Agrarian Change 1640-1750,* Cambridge University Press, Cambridge.

Thirsk, J. and Imray, J. M. (eds.) (1958). *Suffolk Farming in the Nineteenth Century,* Suffolk Record Society, Woodbridge.

Thomasson, A. J. (1975). (ed.). *Soils and Field Drainage,* Technical Monograph No. 7., Soil Survey of England and Wales, Harpenden.

Thomlinson, J. (1882). *The Level of Hatfield Chase,* Doncaster.

Thompson, F. M. L. (1968). The second agricultural revolution, *Economic History Review* **21**, 62-77.

Thompson, F. M. L. (1981). Free trade and the land. In Mingay, G. (ed.) *The Victorian Countryside*, Routledge and Kegan Paul, London, Vol. 1, 103-17.

Todd, M. (1989). Villa and fundus. In Branigan, K. and Miles, D. (eds). *The Economies of Romano-British Villas*, Sheffield University Department of Archaeology and Prehistory, Sheffield, 4-20.

Trafford, B. D. (1970). The history of field drainage, *Journal of the Royal Agricultural Society of England*, **131**, 129-52.

Trafford, B. D. (1985). 'A future for drainage. *Proceedings of the National Agricultural Conference*, Royal Agricultural Society of England/ADAS, National Agricultural Conference, Stoneleigh.

Tubbs, C. R. (1978). An ecological appraisal of the Itchen Valley flood-plain, *Proceedings of the Hampshire Field Club Archaeological Society*, **34**, 5-22.

Tyacke, S. (ed.) (1983). *English Mapmaking 1500-1650*, British Library, London.

Van Beers, W. F. J. (1963). *The Auger Hole Method. A field measurement of the hydraulic conductivity of soil below the water table*, Wageningen.

Vancouver, C. (1808). *General View of the Agriculture of Devon*, London.

Vancouver, C. (1810). *General View of the Agriculture of Hampshire*, London.

Van de Noort, R. and Ellis, S. (eds.) (1997). *Wetland Heritage of the Humberhead Levels: An Archaeological Survey*, University of Hull, Hull.

Van Rijn, P. (1995). Culverts and questions, *NewsWARP: the newsletter of the Wetland Archaeological Research Project* **17**, 37-9.

Van Zeist, W. (1974). Palaeobotanical studies of settlement sites in the coastal area of the Netherlands, *Palaeohistoria* **16**, 223-371.

Vollans, E. (1988). New Romney and the 'river of Newenden' in the later Middle Ages. In Eddison, J. and Green, C. (eds) 1988, 128-41.

Vollans, E. (1995). Medieval salt-making and the innng of the tidal marshes at Belgar, Lydd. In Eddison, J. (ed.) 1995, 118-26.

Wacher, J. (1995). *The Towns of Roman Britain*, 2nd edn, Batsford, London.

Wade Martins, S. (1980). *A Great Estate at Work*, Cambridge University Press, Cambridge.

Wade Martins, S. (1995). *Farms and Fields*, Batsford, London.

Wade Martins, S. and Williamson, T. (1994). Floated water-meadows in Norfolk: a misplaced innovation, *Agricultural History Review* **42**, 20-37.

Wade Martins, S. and Williamson, T. (1995). *The Farming Journal of Randall Burroughes (1794-1799)*, Norfolk Record Society, Norwich.

Wade Martins, S. and Williamson, T. (1997). Labour and improvement: agricultural change in East Anglia c. 1750-1870, *Labour History Review* **62** (3), 275-95.

Wade Martins, S. and Williamson, T. (1999). *Roots of Change: Landscape and Agriculture in East Anglia, 1700-1870*, British Agricultural History Society Monograph, London.

Waller, M. (1994). *The Fenland Project, Number 9: Flandrian Environmental*

Change in Fenland, East Anglian Archaeology 78.

Ward, R. C. and Robinson, M. (1990). *Principles of Hydrology*, 3rd edn, McGraw Hill, London.

Wessex Water Authority (1979). *Somerset Local Land Drainage District Land Drainage Survey Report*, Bridgewater, Somerset.

Wheeler, B. D. and Shaw, S. C. (1987). *Comparative Survey of Habitat Conditions and Management Characteristics of Herbaceous Rich-fen Vegetation Types*, Nature Conservancy Council, Peterborough.

White, R. E. (1979). *Principles and Practice of Soil Science*, 3rd edn, Blackwell Science, Oxford.

White, S. K., Cook, H. F. and Garraway, J. L. (1997). Watercourse protection in marshlands, *SEESOIL* **12**, 39–56.

Whitney, K. P. (1989). Development of the Kentish marshes in the after math of the Norman Conquest, *Archaeologia Cantiana* **107**, 29–50.

Whitwell, B. (1988). Late Roman settlement on the Humber and Anglian beginnings. In *Recent Research in Roman Yorkshire* (eds. Price, J. and Wilson, P. R.), British Archaeological Reports British Series **193**, 49–78.

Wigley, T. M. L., Ingram, M. J. and Farmer, G. (eds.) (1981). *Climate and Change*, Cambridge University Press, Cambridge.

Wilkinson, T. J. and Murphy, P. L. (1995). *The Archaeology of the Essex Coast, Volume 1: The Hullbridge Survey. East Anglian Archaeology* 71.

Williams, D. H. (1984). *The Welsh Cistercians*, Griffin Press, Pontypool.

Williams, M. (1970). *The Draining of the Somerset Levels*, Cambridge University Press, Cambridge.

Williams, M. (1982). Marshland and waste. In *The English Medieval Landscape* (ed. Cantor, L.), Croom Helm, London, 87–125.

Williams, M. (1989). Historical geography and the concept of landscape, *Journal of Historical Geography* **15**, 1, 92–104.

Williams, S. R. (1978). Note in *Cheshire Archaeological Bulletin* **6**, 12–17.

Williamson, T. (1997). *The Norfolk Broads. A Landscape History*, Manchester University Press, Manchester.

Wisdom, A. S. (1979). *The Law of Rivers and Watercourses*, 4th edn., Shaw and Sons, London.

Wood, E. B. (ed.), (1897). *Rowland Vaughan, His Booke*, London.

Wordie, J. R. (1985). The South: Oxfordshire, Buckinghamshire, Berkshire, Wiltshire, and Hampshire. In Thirsk, J. (ed.) 1985a, 51–77.

Worlidge, J. (1669). *Systema Agriculturae*, London.

Wright, Rev. T. (1789). *An Account of the Advantages and Methods of Watering Meadows by Art*, Cirencester.

Wright, T. (1792). *An Account of the Advantages of Watering Meadows by Art, as Practised in the County of Gloucestershire*, London.

Wright, Rev. T. (1799). *The Art of Floating Land in Gloucestershire*, London.

Young, A. (1769). *Six Weeks Tour through the Southern Counties*, London.

References

Young, A. (1771). *The Farmer's Tour through the East of England*, London.

Young, A. (1786). Minutes relating to the dairy farms of High Suffolk, *Annals of Agriculture* **27**, 193-224.

Young, A. (1793). A tour through Sussex. *Annals of Agriculture* **22**, 44.

Young, A. (1795). On some watered meadows in Hampshire. *Annals of Agriculture* **23**, 264-8.

Young, A. (1797). Untitled comments in *Annals of Agriculture* **16**, 127.

Young, A. (1799). *General view of the Agriculture of the County of Lincolnshire*, London.

Young, A. (1804a). *General View of the Agriculture of the County of Norfolk*, London.

Young, A. (1804b). *General View of the Agriculture of the County of Hertfordshire*, London

Young, A. (1807). *General View of the Agriculture of the County of Essex*, 2 vols, London.

Young, A. (1813a). *General View of the Agriculture of the County of Suffolk*, London.

Young, A. (1813b). *General View of the Agriculture of the County of Buckinghamshire*, London.

Young, Rev. A. (1808). *General View of the Agriculture of the County of Sussex*, London.

Index

Entries in *italics* refer to figures

'acid sulphate' soils, 80, 81-3
   lime requirements of, 80
'ADAS', 227
'Adventurers', 10, 147
aeration, 15-16
aerobic respiration, 16, 73
agricultural imports, 12, *154*
agricultural regionalisation, 9, 11
'agricultural revolution', 11, 12
Agriculture Act 1937, 234
Agriculture Act 1947, 235
Agriculture Bill 1957, 235
alluvium, alluvial soils, 2, 3
Amberley Wild Brooks, 185, 237-9
anaerobic conditions, 17-18, 73-4
Anglo-Saxon period *see* Saxons
animal production, 3, 101, 109, 113
arable production, 3, 4, 10, 11, 13
   conversion from pasture, 51
   and drainage, 25-7, 28-40, 41
   primary wetlands, 103
archaeological methods in wetlands, 106, 116
archaeological resource, 4, 14, 101-2
aristocratic 'improvers', 198, 207
Arun, River
   see also Amberley Wild Brooks
Avon Navigation
   see Salisbury

Avon, River (Salisbury), 165, 183, 194
   see also Salisbury

base-flow (of rivers), 163
Bedford Rivers, 144-6, *145*, 212
Bedwork water meadows, 160-3, *166-7*, *181*, *200*, *201*, *202*
   microclimates, 173, *174*
biodiversity, 13, 177, 210-26
Black Death, 138-9
Blith, Walter, 47
Board of Agriculture, 11, 56, 187, 198, 228, 229, 230
Boswell, George, 187, 189-90, *191*
Bronze Age, 14
Burwell Fen, drainage of, 88, *89*, 150
'bush drainage' *see* underdrainage
butene, 18
Britford SSSI (near Salisbury), 165-78, *166-7*, *168*

Caird, James, 43-4, 54, 70
Cambridgeshire, reclamation in, 141-6, 149, 150
Car Dyke, 93, 118, 151
carbon dioxide, 15, 18
Castle Acre (Norfolk), 199-203
Catchment Boards/River Boards,

156, 232, 233, 234, 236
'catchwater drains', 93, 117–18
catchwork, water meadows, 159–60, 160–1, 179, 190–3, 199, 205
cattle, on water meadows, 164
chalklands/chalk streams *see* Downlands
Charles I, 147, 211
claylands/clay soils, 9, 11, 21–2, 34, 36, 40, 42, 47, 48, 49, 50, 52, 60, 62
climate and soil water, 18–20
  see also irrigation of grassland
climatic factors, 2, 18–20, 151–2
Clipstone Park (Duke of Portland's Water Meadows), 159–60, 160–1, 203, 208
clover, 51, 207
Cobbett, William, 186
crop rotation, 53–4
Coke, Thomas William, 198
Commissions of Sewers/Bill of Sewers, 8, 9, 139, 146, 148, 150, 228, 231
common land/common pasture, 137, 149
Countryside Stewardship scheme, 14, 164
County War Agricultural Executive Committees (CWAECs), 233

'darlands', 40
Davis, Thomas, 186, 187, 188
demography/demographic trends, 1, 5, 6
  at Domesday, 7
  late eighteenth century, 49
  late medieval, 9
  post-medieval, 49, 153
  Roman, 7
  Saxon, 7

denitrification, 17
Denton, J. Bailey, 54–5
diesel-powered pumps, 88, 95, 99, 100, 151
dissolved load, 171–3
ditch maintenance, 41
Domesday Book, 7, 136
Downlands, 5, 22, 42, 179, 182, 183, 185, 188, 195
drain depth, 25, 63
drain spacing, 25
drainage, 20–1, 54, 141–2
  benefits from, 15, 26–7, 52, 53–4, 65
  and conservation, 227–43
  economics of, 27, 152–6, 234
  field-scale, 2, 10
  investment, 68–71
  nineteenth century, 60–1, 53–72
  regional, 7, 9, 10, 84, 92, 94, 210
  of ridge and furrow, 33
  Romans, 121
  surface, 28–40, 41–3, 67
  technology, 117–20, 142–3
  tenant involvement, 70–1
  in water meadows, 162
  see also diesel; flood defence; gravity; steam; underdrainage; windmills
Drainage Boards, 228, 230–1
drainage density, 85–8
'drowners', 163
Duke of Portland's Meadows *see* Clipstone Park

East Anglian Fenland, 3, 10, 11, 12, 122, 123
  conservation designations in, 212–13, 236
  drainage of siltlands, 40
  early drainage of, 22
  ecology of, 210–36

267

economy to the seventeenth
  century, 10-11, 149, 210-12
 in medieval period, 125, 127-30,
  132-5, 137-9
 in post-medieval period, 144-7,
  152, 153, 154
 regional drainage of, 93-4
 in Roman period, 102-3, 104-5,
  106, 109, 113, 114-17, 119,
  120-1
 seventeenth-century drainage, 88
East of England, 11, 22, 36, 40-2,
  44, 47-52, 55, 58, 59, 69, 70,
  204-8, 227
Ebble, River, 184
economics of drainage, 27, 152-6
economy, national
 eighteenth century, 49
 medieval, 5
 nineteenth century, 154, 203-4,
  243
ecosystem change, 3, 5, 210-26
electrically-driven pumps, 95, 100
embankments, 2, 7, 10, 93, 108,
  125, 132, 137, 143
enclosure, 10, 11, 28, 34, 39, 40, 49,
  67, 137, 148
 ancient enclosed communities, 51
English agriculture, characteristics of,
  2-4
Environmentally Sensitive Areas, 164
Essex Marshes, 13, 102
ethylene, 18
European (Economic)
  Community/Union, 13, 240
eutrophication, 13
evaporation/evapotranspiration, 15,
  20
 'potential transpiration', 19
Exmoor, 19, 83, 192, 193, 203

fallow fields, 31

Fen communities
 animals, 225-6
 humans, 149, 210-12
 plants, 212-25
Fenland/'The Fens' see East Anglian
  Fenland
fertilisers (artificial), 4, 12, 164, 193
feudal system, 8
'field capacity', 20, 92
fields (as collection of furlongs), 31
First World War, 229
Flag Fen, 14, 158
Flavian Emperors, 114
'floating' of water meadows see
  irrigation of grassland
flood defence/flooding, 4, 6, 7, 8,
  19, 26, 75, 84, 101, 125, 140,
  143, 152, 166
flood meadows, 31, 213, 214-15
Floriacensis, Abbot, 210
Folkingham, William, 42
'freeboard drainage', 16, 88, 90
Frome, River, 194
frosts (limitation to growth of grass)
  205-6
'furlongs', 31
furrows and drainage, 32, 33

General Drainage Act 1600, 10, 147
'gleying', 17
'gouts' (tidal doors), 120
grant in aid
 for irrigation, 198
 for underdrainage, 156, 228, 233,
  237
gravity drainage, 88, 100
grazing pastures, 31, 158
Great Ouse, River/Ouse Drainage
  Board 133, 229-30
'green history', 1
'green revolution', 4
'gripes' (possible Roman origin), 119

Index

Gwent Levels, 103, 120, 122, 130-1, 135

habitat loss, 10, 13, 211, 213, 226, 227, 236, 238
Hadrian, Emperor, 113-15
Halvergate Marshes, 3, 13, 85
Hardy, Thomas, 187
'heads' (soil banks), 31
Herefordshire, irrigation in, 179-80
'high farming' period, 12, 203-4, 243
'Hooghoudt equation', 25
House of Lords Select Committee on Science and Technology, 241
Humber, River, 122, 132, 153, 155
'hydraulic civilisations' 2, 15
hydrology
    of reclaimed land 84-100, 143-6
    of water meadows 159-71

Inclosure Commissions, 67
Indigenous Technical Knowledge (ITK), 4
industrial development, 4
infiltration rates in water meadows, 169
institutions for water management, 6, 7
    see also Catchment Boards
Internal Drainage Boards (IDBs), 148, 232, 233, 236, 239
irrigation of grassland, 1, 3, 26-7
    duration of, 167
    'floating upwards', 158
    for hay crop, 197
    microclimatic factors, 173, 205-6
    origins, 157-9, 163-4
    prizes for, 198-9
    uncommon counties for, 196
    see also sub-irrigation; water meadows
Iron Age settlement, 101
iron compounds, 73-4, 80
Itchen, River, 182, 188

James I, 147
jarosite, 80
'jurats', 7

Kennet, River, 182, 185, 195
Kent, Nathaniel, 41, 42, 43, 46, 205
Knight, John and Frederick Winn, 192-3

Lambourn, River, 185
Land Drainage Act 1861, 228
Land Drainage Act 1918, 228
Land Drainage Act 1930, 155-6, 232-5
'lands', 30, 32, *32*
landscape, impact of agriculture upon, 3
Lavant, 185
Lincolnshire, 103, 115-16, 123, 127, 133, 136, 142, 149, 153, 204, 205
'Little Ice Age', 152

manorial courts, 8, 10, 30, 40
    as 'multifunctional agencies', 8
manure, 3
Marshall, William, 45, 49, 56-7
marshland drainage, 4, 23, 84-8, *87*, 91, 108-11, 117-20, 123-5, 133, 142-6
meadow irrigation *see* irrigation of grassland; water meadows
meadows, definitions, 157-8
medieval period, 3, 5, 10, 24, 28-40, 122-40
Meon, River, 182, 194
methane/methanogenesis, 17-18

269

Middle Ages *see* medieval period
'Midland System', 36
Midlands, 4, 6, 10, 11, 14, 21, 22, 24, 28, 34, 42, 49-50, 55, 58
Ministry of Agriculture, Fisheries and Food (MAFF), 227-8, 228-9, 233, 239-42
mole drainage, 25, 59, 63, 233
monasteries, 8, 125, 135-6, 138, 210
mottling (in soil), 73, 175

Napoleonic Wars, 11, 42, 49, 154, 195, 201, 204
Nar, River, 199
National Nature Reserves, 212, 225, 235, 236
National Parks Commission, 228
National Vegetation Classification (NVC), 173-7, 213-24
Nature Conservancy (Council), 228, 235, 236-9, 240, 241
Nene, River, 133
  Washes, 236
Neolithic period, 3
nitrogen compounds, 16, 73, 74
  nitrate 16, 26, 172-3
Norfolk
  floating in, 199-203
  ridge and furrow in, 39
Norfolk Broads/Broadland, 9, 11, 12, 95, 100, 131
Norfolk Marshlands *see* East Anglian Fenlands
Norman Conquest, 127
North Kent Marshes, 13, 102
  Environmentally Sensitive Area, 241
Northern counties, 30, 37, 132, 137, 141, 153, 154, 198, 204, 229, 231
  ridge and furrow, 36, 37, 55

open field agriculture, 6, 28, 30, 36
  and poor drainage, 41
Otmoor, 123
Ouse (Sussex), River, 131
Ouse (Yorkshire), River, 229
Ouse (Bedfordshire) Catchment Board, 232
Ouse Washes, 212
oxygen/oxidation, 74, 172
  see also reduction-oxidation

pasture, conversion from arable, 49
peat/peatlands, 2, 3, 12, 23, 75, 88, 89-90, 92, 132-4, 137, 153, 210-26
  peat cutting, 103, 137
  see also soil, shrinkage
Pevensey Levels, 122, 123, 131, 138, 139
Piddle (Dorset), River, 180
plant communities, 173, 177, 212-25
ploughs/ploughing, 31, 32, 34
  and surface form, 42
  two-way, 39-40
polders/'empoldering', 85
Polyolbien, 211
population trends *see* demography
precipitation, 1, 18-20
pumps/pumping technology, 94-100, 119, 151
  see also 'scoop-wheels'
Pusey, Philip, 46, 64, 203, 205, 208

Quarme, River, 192

rainfall *see* precipitation
raised mire/bog, 75, 224
recession (agricultural), 154-6, 194, 204
reclaimed land, 2, 7-10, 101
  see also wetland

Index

reclamation, definition, 142
  stages of, 85-8
Redgrave and Lopham Fens, 236
reduction-oxidation reactions in soils, 16-18, 73-5
'reens' or 'rhynes', 119-120
Reformation, 146
respiration processes, 74
Rhee Wall, 134
ridge and furrow, 10, 24, 29, 32, 33, *35*, 28-40, 119
  dating of, 38
  'narrow rig', 30
  ploughing cause of, 32-3, 42
  post-medieval, 41-3
  regions where rare, 38-40
  straight, 28-9
River Boards Act 1948, 234
  see also Catchment Boards
river navigation, 8
rivers *see under individual names*
'roddons', 103, 116
Romney Marsh region, 7, 8, 22, 102, 122, 131, 134, 136
  habitat change, 13
  historic land use, 85, 138
  landscape, 109
  methods of draining, 94, 100, 125
  soils, 88-90, 123
'Roman Bank', 127
Roman draining-spade, 227
Roman period, 7, 94
  legions, 109-12
  mining, 113
  reclamation, 101-21, *104-5, 107, 110-11*
  villas, 112
root crops, 2, 51, 206
Rother, River, 185
Royal Agricultural Society, 203
  Journal of the, 60

Royal Military Canal, 94
Royal Society, 182
Royal Society for the Protection of Birds, 240

Salisbury, 19, 165, 172, 182, 184, 206
  Avon Navigation, 165
salt production, 101, 102, 103, 109, 112, 117
saltmarsh *see* wetland
saturated hydraulic conductivity, 'permeability', 25, 170
Saxons, Anglo-Saxon period, 7, 127, 138
'scoop-wheels', 94
sea levels, 2, 75, 106
Second World War, 2, 13, 22, 27, 50, 195, 233, 242
sediments
  entering water meadows, 170-1
  loading of watercourses, 4
  of marshlands, 75-6, 101, 102, 123
  in 'warping', 158
Select Committee into the State of Agriculture, 63
settlement in wetlands, 113-17, 124-5, 127, 130
Severn Estuary, 122-3, 139 *see also* Gwent Levels; Somerset Levels; Wentlooge Levels
Sewers, Bill of, 9 *see also* Commissions
'sheep-corn system', 5-6, 163, 186, 204
Sherborne Estate, 195
Shoard, Marion, 242
Sites of Special Scientific Interest (SSSI), 89, 165, 235, 236, 239, 240
Smith, Robert, 193

INDEX

Smith, William, 197, 199, 205-6
social factors in drainage, 149-51
soil, 16-18
    changes after reclamation, 76, 79-83
    mapping, 21-3, *23*, *56*, 75-6
    of managed wetland, 80-3, 88-90
    of natural wetland, 76-9
    and 'ridging', 33, 34, 39
    and 'ripening', 76
    series and drainage requirements, 21-3
    (peat) shrinkage, 88, 100, 144, 153, 213
    'subsoiling' in water meadows, 162
    texture, 20, *56*, *62*, *69*, 92
    of upland wetland, 83
    water deficit, 92
    water profiles, 20-2, *21*, 167-9, *169*
    water retention curves, *92*, *93*
Somerset Levels (siltlands) and moors (peatlands), 8, 14, 84, 85, 102, 122, 141, 146, 148, 150, 227
    drainage of, 88, 100, 106
    medieval period, 130, 132-3, 135, 137
    peat, 22, 88, 132-3
    Roman period, *107*, 108-9, *108*, 119, 120-1
Southern Water Authority, 237-8
steam-pumping technology, 88, 95, 100, 151
Stiffkey, River, 199
'stitching', 42
strips *see* 'lands'; ridge and furrow
sub-irrigation, 85, 92
sulphur compounds, 73-4
sustainability of water meadows, 177

'sustainable development', 4
Swale, 240

Tadham Moor, 88
technology transfer and innovation, 5, 6, 189-90, 207
Test, River, 182, 184, 195
textile industry, 49
Thames Estuary marshlands, 13, 123, 131, 137, 141
'tile drainage' *see* underdrainage
Tithe Commutation Act 1836 Districts, 62
trafficability, 15, 26
turbary, 10, 88, 137

underdrainage, 5, 11, 12, 24, 27, 53-72, 90
    'bush', 43-51, *45*, 59
    cost of, 46, 50, 63
    finance, of 65-8
    meadows, 46, 162
    nineteenth century, 53-72, *57*, *61-2*, *69*
    pebbles/boulders, use of, 45, 59
    post-medieval, 43-6
    'tile'/pipe, 43, 44, 48, 50, 63-5
    and yield, 52
upland, 36
urban development, 4, 121

Vancouver, Charles, 188, 192
Vaughan, Rowland, 179-80
vegetation communities, 173-7, 210-25
Vermuyden Cornelius, 153, 211
villas and Roman estates, 112, 116-17
volatile fatty acids, 18

wages, 5

Walland Marsh *see* Romney Marsh
warping, 158
waste, 10
Water Act 1973, 237
water meadows, 3, 4
  aesthetic value of, 208
  cost of construction, 162, 183–4, 187–8
  decline of, 13, 164, 193–5
  enhanced grass production, 164–5, 180, 185
  and geology, 162
  infrastructure, 159–63, 179, 183, 195
  management, 163–5, 185–7
  in north and east, 196–209
  operation of, 157–209
  plant communities in, 173–7
  profitability, 188
  in south and west, 179–95
  see also irrigation
watermills, 205
watertable, 24–6, *24*, *32*, 73, 90–1
Weald, 38
Wentlooge Level, 102, 108–9, *110–11*, 130, 135
West of England/Wessex, 11, 157, 162–78, 179–95, 196, 201, 204
wetlands, 2, 7, 9, 13, 14, *124*
  economy of, 101, 102–6
  hydrology, 84–100
  medieval reclamation, *126*, *128–9*
  plant communities and flooding, 173–7, 212–25
  reclamation, 84–100, 122–40, *126*, 141–56
  soils, 73–83
  stages in reclamation, 84–8, 126
  see also drainage; soils
Wey, River, 182, 194
Whittlesey Mere, 144, 213, 224
windmill technology, 94–5, *96–9*, 100, 211–12, *212*
Windrush, River, 195
Wolds (Yorkshire), 5, 37
Wright, Revd Thomas, 190
Wylye River, 182

'yardland', 30
Yield Benefit Index, 26
Young, Arthur, FRS, 43, 45, 47–8, 58–9, 197, 207
Young, Revd Arthur, 58, 188